Microbiology

49 Science Fair Projects

Other books in the series

Insect Biology: 49 Science Fair Projects
H. Steven Dashefsky

Botany: 49 Science Fair Projects
Robert L. Bonnet and G. Daniel Keen

Botany: 49 More Science Fair Projects
Robert L. Bonnet and G. Daniel Keen

Computers: 49 Science Fair Projects
Robert L. Bonnet and G. Daniel Keen

Earth Science: 49 Science Fair Projects
Robert L. Bonnet and G. Daniel Keen

Environmental Science: 49 Science Fair Projects
Robert L. Bonnet and G. Daniel Keen

Space and Astronomy: 49 Science Fair Projects
Robert L. Bonnet and G. Daniel Keen

Microbiology

49 Science Fair Projects

H. Steven Dashefsky

Illustrations by Debra Ellinger

TAB Books

Division of McGraw-Hill, Inc.

New York San Francisco Washington, D.C. Auckland Bogotá
Caracas Lisbon London Madrid Mexico City Milan
Montreal New Delhi San Juan Singapore
Sydney Tokyo Toronto

Published by TAB Books, a division of McGraw-Hill, Inc.

pbk 1 2 3 4 5 6 7 8 9 0 FGR/FGR 9 9 8 7 6 5 4
hc 1 2 3 4 5 6 7 8 9 0 FGR/FGR 9 9 8 7 6 5 4

Library of Congress Cataloging-in-Publication Data

Dashefsky, H. Steve.
Microbiology : 49 science fair projects / by H. Steven Dashefsky.
p. cm.
Includes index.
ISBN 0-07-015659-X ISBN 0-07-015660-3 (pbk.)
1. Microbiology—Juvenile literature. 2. Biology projects—Juvenile literature. [1. Microbiology. 2. Science projects.] I. Title.
QR57.D37 1994 93-48718
576'.078--dc20 CIP
AC

Acquisitions editor: Kim Tabor
Editorial team: Joanne Slike, Executive Editor
Annette Testa, Book Editor
Joann Woy, Indexer
Production team: Katherine G. Brown, Director
Ollie Harmon, Coding
Tina M. Sourbier, Coding
Jan Fisher, Desktop Operator
Lorie L.White, Proofreading
Design team: Jaclyn J. Boone, Designer
Brian Allison, Associate Designer
Cover design and illustration by: Holberg Design, York, Pa.
Back cover copy written by: Michael Crowner
Technical reviewer: Paul J. Hummer, Jr., Adjunct Professor, Education Department, Hood College, Frederick, Md.

0156603
SFP

Acknowledgments

The author would like to thank Dr. Paula J.S. Martin for her technical assistance in performing the projects in this book and with the preparation of the manuscript.

Disclaimer

Adult supervision is required when working on these projects. All the projects within this book assume your teacher or some other knowledgeable adult advisor will be assisting you throughout the project. No responsibility is implied or taken for anyone who sustains injuries as a result of using the materials or ideas, or performing the procedures put forth in this book.

Use proper equipment (gloves, forceps, safety glasses) and take other safety precautions. Read and follow the manufacturer's instructions when using equipment and chemicals. Use chemicals, dry ice, boiling water, flames, or any heating elements with extra care. Wash hands after project work is done. Taste nothing. Tie up loose hair and clothing. Follow step-by-step procedures and avoid shortcuts. Never work alone. Additional safety precautions are mentioned throughout the text and in the section "Safety in the Microbiology Laboratory." If you use common sense and make safety a first consideration, you will create safe, fun, educational, and rewarding projects.

Contents

APPENDICES

Safety in the microbiology laboratory

All the experiments in this book require adult supervision. This should be your teacher or some other adult advisor who is knowledgeable about microbiology laboratory procedures, including safety and disposal techniques.

Be conscious of possible contaminants, especially on your hands. Clean up your work surface, especially if you had a spill of any kind. Wash your hands frequently. Never eat or drink in a laboratory. Wear protective clothing (lab coats, goggles and gloves); if you contaminate your clothing, remove it and wash it. Good sterile technique leads to good, safe habits.

Some of these experiments will produce fungal (mold) growths. Fungi produce spores that disperse in the air. A plate with a fungal growth, if opened, will contaminate the air with high numbers of spores. High spore densities in the air can cause respiratory problems for people with allergies or asthma. If possible, do not open plates with mold growths. If you must, open them slowly and carefully, in a room with very little air movement. Work with your teacher throughout all projects.

Some microbes are pathogenic. None of the experiments in this book uses pathogenic microbes. However, whenever you are isolating microbes from the environment (and from your mouth or from a cough), you might culture some pathogens. It is extremely important that you dispose of all cultures in a safe manner. If you have cultures on agar plates, be sure to seal the plates with adhesive tape and wrap the plates in a biohazard bag (Fig. I-1) and have your teacher handle their proper disposal.

The best way to dispose of cultures is to autoclave them before disposal to be sure all the microbes are dead and will not contaminate anything (Fig. I-2). This is especially important for plates with fungal growth.

Before beginning, the student and the adult sponsor should read and review the entire project. The adult sponsor should determine beforehand which elements of the experiment require adult supervision. In addition, the caution icon (the triangle with an exclamation point inside), as shown in the margin to the left, will appear throughout the text to indicate where extra caution must be taken.

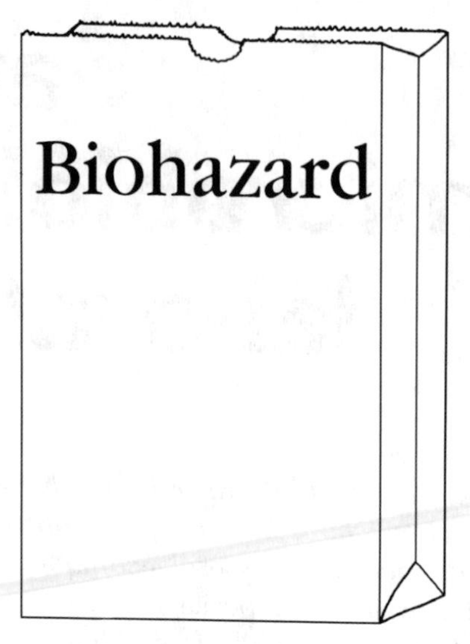

I-1
Your teacher or other adult advisor must direct the proper disposal of all cultures and materials used for these projects.

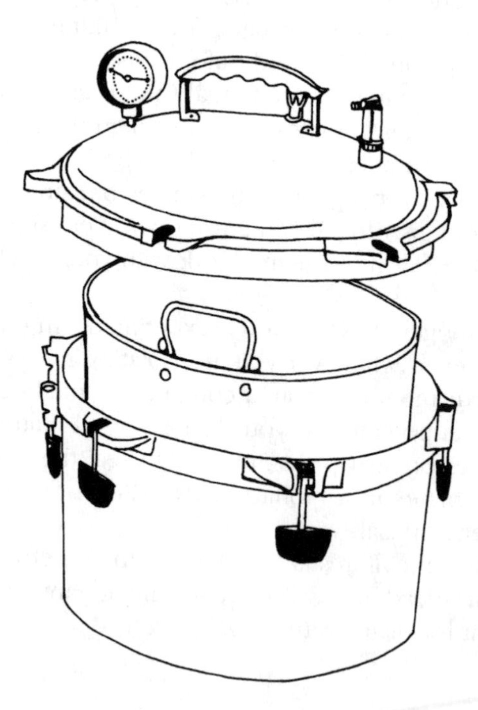

I-2
An autoclave uses steam and pressure to sterilize equipment.

How to use this book

Each of the 49 projects in this book has an Overview (the first few paragraphs of each project), Materials, Procedures, Conclusion, and Going Further section. Each experiment gives you the step-by-step instructions, but leaves you to draw your own conclusions from the data you collect.

Overview

The Overview section gives background information about the topic, explains the purpose of the experiment, and poses questions. These questions will help you develop a hypothesis for your experiment. Developing a hypothesis and the scientific method are discussed in more detail in chapter 2 "Getting Started."

Materials

The Materials section lists everything needed to perform the experiment, but you can improvise when necessary. The following materials might not be listed, but should be available for any of the projects: a pad and pencil for note taking, Scotch tape, scissors, and water.

Procedures

The Procedures section gives step-by-step instructions on how to perform the experiment and suggests how to collect data. *Be sure to read through this entire section before undertaking any project.*

Conclusions

The Conclusions section doesn't draw any conclusion for you. Instead, it asks questions to help you interpret the data and reach your own conclusions.

Going further

The Going Further section is an important part of every project. It lists many ways for you to continue researching the topic beyond the original experiment.

Suggestions are given on what to read and what additional experimentation can be performed. Performing some of these suggestions will assure you that the topic has been thoroughly covered and show you how to broaden the scope of the project. The best way to ensure an interesting and fully developed project is to include one or more of the suggestions from the Going Further section.

This book is designed for sixth- to ninth-grade students, but it can be used by older students. High school-age students should include the Going Further suggestions in each experiment to broaden its scope and depth. Combining related projects is another excellent way to adapt these projects for older students.

The projects in this book

The projects in this book have been divided into eight sections (Part One, Microbes & Microbiology, offers an introduction to microbiology and how to get started in this field.). Each group after the initial section begins with an introduction that explains the importance of the experiments included. These sections are as follows: Part Two "Microbes of Humans & Animals" investigates microbes that live on us, plus those that inhabit our pets. Projects in this section also explore our favorite habitat—the backyard.

Part Three "Microbes of Food" looks into how microbes can both destroy and produce foods that we eat. Projects also investigate how we try to preserve our foods. Part Four "The History & Tools of Microbiology" compares the past and present methods of studying these organisms and teaches us a great deal about the fundamentals of microbiology.

Part Five "The World of Bacteria" investigates the largest of the five groups of microbes. Bacteria are believed to be the first form of life to have appeared on our planet. Part Six "Other Microorganisms" includes projects about the other four groups of microbes: algae, fungi, protozoans, and viruses. Each of these groups plays an important and distinct role on our planet.

Part Seven "Control of Microorganisms" looks at how we fight off microbes, whether in our wounds, on our foods, or in our bathrooms. Part Eight "Microbes & Disease" investigates how they cause disease not only in humans, but in other organisms, and shows how these microbe-produced diseases can be used to our benefit.

Finally, Part Nine "Microbes & the Environment" explores the vital role that microbes play in almost every ecosystem. Everyone is talking about recycling, but microbes are the ultimate and original recyclers. Whether we are aware of it, microbes live with us; they are in the air we breath, the water we drink, and the food we eat.

Part one

Microbes & microbiology

Organisms that are too small to be seen with the naked eye are called *microbes.* Too often, these microbes are thought of as simply "germs" that make us ill and put us in bed with the flu. In truth, however, microbes are one of the most important groups of living organisms found on our planet. A teaspoon of rich soil can contain billions of bacteria, millions of fungi, hundreds of thousands of algae and protozoans—all tiny invisible microbes.

Microbes are found all over the world. They are in the soil, in the air, at the bottom of the sea, in hot springs, and almost any other place on our planet. Microbes can be divided into five major groups: bacteria, algae, fungi, protists, and viruses. Some of these organisms are *producers*, meaning they make their own food during photosynthesis, just as green plants do. Others are *consumers*, meaning they must "eat" their food. Some of them are predators that actually attack and devour other microbes. Still others are *decomposers*, meaning they feed on dead, decaying organisms. These microbes play an important role in helping to recycle nutrients so they can be used by another generation of living plants.

1

An introduction to microbiology

Learning the basics

Early microbiologists found a complex, teeming, tiny world, but the size of the microbes made it difficult to study them. For example, without high magnification, one bacteria looked pretty much like another. Similarly, cells are often too small to study individually, so groups of them, called *colonies*, are studied. When we grow these colonies in an artificial environment, such as a petri dish, they are called *cultures*. A *pure culture* of microbes means the culture contains only one type of microbe and that it has not been contaminated with any other type of microbe. To create pure cultures, scientists had to develop *sterile techniques*. These techniques include methods used to sterilize glassware, inoculate loops, and spreading rods, and a material for the microbes to grow on called a *growth media*.

Microbiologists had to develop a growth media that would nourish the microbes and provide a solid surface on which they could live and reproduce. *Agar*, a compound that gels to a solid after boiling, is perfect for growing (culturing) colonies of microbes. Culturing pure colonies on a solid surface lets microbiologists see characteristics of the microbe, such as the shape, size, and color of colonies. Growth media and sterile technique are required tools to see and study the tiny, fascinating world of microbes. The most important tool that has aided in the study of these microbes is, of course, the *microscope*.

AN INTRODUCTION TO MICROSCOPES

Microscopes have been crucial to the development of microbiology. They allow us to see the smallest living creatures. Microscopes magnify objects, letting us see things that are smaller than our eyes would permit. Our ability to see is limited by *resolving power*. This means it is limited by the ability to see two separate objects that are extremely close to one another. If two objects appear to be one, then we have not *resolved* them. Microscopes give us much greater resolving power, or *resolution*, than our eyes.

The most common type of microscope is called the *compound microscope* (see Fig. 1-1). This microscope has one or two eyepieces, called *oculars.* The *eyepiece* is the part of the microscope that you look through to see the magnified image of the object. The eyepiece contains a lens that magnifies the microbe 10 to 12 times. If the scope has two eyepieces, it is called a *binocular scope.* If the scope has only one eyepiece, then it is called a *monocular scope.* (Don't confuse a binocular microscope with a *stereoscope,* also called a dissecting scope, which is mentioned later in this chapter.)

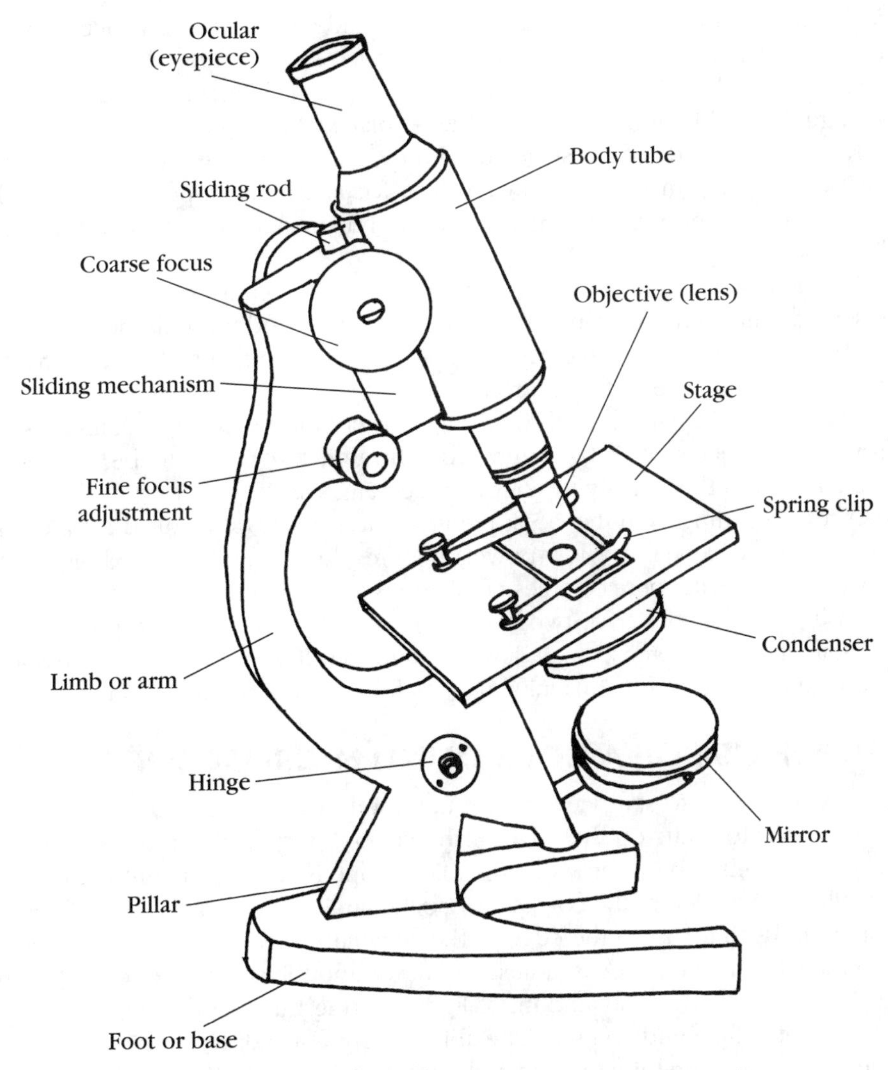

1-1 Become familiar with the various parts of a typical compound microscope.

In addition to the ocular, a *compound microscope* contains at least one objective lens. These lenses magnify the power of the eyepiece. Most school scopes have a low-power objective (10× magnification), a high-power objective (40×), and some have an oil-immersion objective (100×). Oil immersion is discussed in more detail in the next chapter. If you have a 10× ocular lens and a 40× objective, your total magnification is 400×. It is always best to start viewing an object at the lowest power, then switch to a higher power, and then go to the highest objective.

Microscopes have a *stage* that holds a microscope slide containing the object to be viewed. A drop of water, a hair, a fish's scale, or any other small object of interest can be placed on a microscope slide, which is then covered with a *coverslip*. The slide is then placed on the microscope stage. The stage has a hole in it through which light can pass. There are usually metal clips attached to the stage that hold the microscope slides in place (see illustration).

Knobs on the side of the stage adjust the distance between the slide and the objective so you can bring the image into focus. Some scopes have a single coarse focusing knob, but others have a second smaller focusing knob for fine adjustments.

Beneath the stage is a *substage*. Here the light from a light source is condensed and directed up through the hole in the stage and through the object being observed. The light source can be sunlight or a lamp that is directed through the substage with a reflecting mirror.

Once the light passes up through the object being viewed, it continues up through the objectives to be magnified—through the ocular to be magnified once again, and then finally to your eye (see Fig. 1-2).

A few experiments in this book require a *stereoscope* (also called a dissecting scope). A stereoscope is a low-power, high-resolution microscope designed to view objects such as entire colonies of microbes growing on a petri dish or the gills of a fish. These scopes have two objectives as well as two oculars, producing a three-dimensional (stereoscopic) view of the object. If you don't have access to a stereoscope, you can use a magnifying glass for most of the projects in this book.

PROPER USE OF A COMPOUND MICROSCOPE

It is always easiest to start viewing through the lowest power (10×) objective. It is also easier to focus on the edge of the coverslip, rather than trying to focus first on a tiny microbe. Even when the microscope is completely unfocused, you can still see the edge of the coverslip. Use the large (coarse) and small (fine) focusing knobs to focus on the edge of the coverslip.

Once the edge is in focus, adjust the mirror and the condenser to get just the right amount of light. The amount of light entering through the stage is important. Too little light and you won't be able to distinguish different structures. Too much light and everything will be a glaring blur. Most microscopes have a *condenser* with which you control the amount of light. Once the light is adjusted

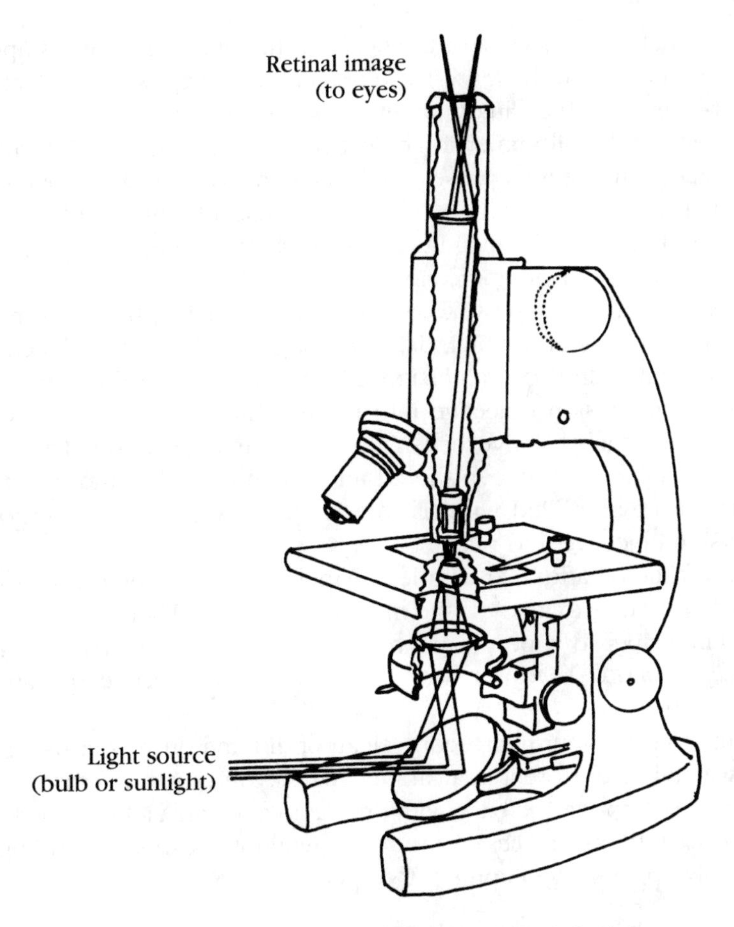

1-2 The best way to understand how a compound microscope works is to follow the path of light.

and the edge of the coverslip is in focus, move the slide on the stage until the microbe is in view. Use the fine adjustment knob, if available, to get the object in perfect focus.

Now that the light is adjusted and the object is in focus, turn the objective wheel to the next higher power, which is usually 40×. The object will still be approximately in focus. You should only use the fine focusing knob, if available, from this point forward. Notice that the higher power objective is longer than the lower power objective and the oil-immersion objective is the longest of them all. There is very little room to move the stage up and down (to focus) with the high-power objectives. There is a danger of coming down too far and cracking the slide, thus damaging the objective. Therefore, try to do most of the focusing at the lower power, where there is room for movement. Don't let the objective lens contact the coverslip.

Do not touch the objective or the ocular with your fingers. It is important to keep most of the dirt off the lens. If you do get them dirty, wipe them clean with lens paper—never with ordinary paper.

You must use the oil-immersion objective to see any details of bacteria cells. With oil immersion, the light passes up through the stage and microbe, and then through oil instead of the air. This allows you to use a far higher objective—usually 100×. resulting in 1000× magnification, (100×10). Some of the projects in this book call for the use of oil immersion.

To switch to the high-power, oil-immersion objective, first be sure to have the object in focus under the high dry (40×) objective. Move the objective out of the way and put a drop of immersion oil on the center of the coverslip. Now move the oil immersion objective into place. Watch this from the side of the stage to be sure the objective doesn't crunched into the slide. There is just a small amount of room between the oil-immersion objective and the slide. This space should now be filled with oil. Always be very careful when moving the oil-immersion objective into place.

Clean off the oil after use with lens paper. Do not get oil on the dry objectives. Use lens paper to take off the majority of the oil. Then use a tiny drop of xylol on lens paper to wipe off any remaining oil. (*Use xylol with caution; it is toxic and flammable.*) Do not use alcohol to clean objectives because it can damage an objective.

Be sure to keep your microscope clean of dirt and dust; after using it, either cover it with a bag or store it in a cabinet. When you carry a microscope, always use two hands. Support the base with one hand and carry it by its neck with the other. It is easy to lose pieces of the scope when it is carried at an angle. They are expensive pieces of equipment to replace!

USING STERILE TECHNIQUE

While performing most of the projects in this book, the most important thing to do is to try not to *contaminant* your work by allowing unwanted microbes to get in. The only way to prevent contamination from ruining your project is to use a good *sterile technique*, also called *aseptic technique*. Sterile technique is described below, but be sure to get your teacher or advisor to help demonstrate to you these techniques when they are needed for your project.

Sterile technique is simply the ability to transfer (*inoculate*) microbes from their source to a sterile environment (called culture media) without allowing any unwanted microbes to contaminate the media. This means using equipment, supplies, and media that are free of microbes (meaning they are sterile). All glassware and utensils must be sterilized before they touch the culture, otherwise the culture will become contaminated with whatever microbe is on the utensils. (See the section below for details on sterilization of glassware and utensils.)

Because microbes can enter a culture from the air, sterile technique also includes keeping culture vials and plates open for the shortest amount of time possible. This means that all opening and closing of sterile containers should be kept to a minimum and done in a room with very little air movement. It also means the proper use of sterile inoculation loops, spreading rods, and pipettes as described later. Don't wave these utensils around in the air while being used because they could become contaminated.

In addition, sterile technique means never setting down vial caps or plate lids on a surface, which usually contains more microbes than the air, and keeping your work space clean. The workspace surface should be disinfected every day with a bleach/water solution (1 part bleach to 100 parts water).

Most importantly, sterile technique means keeping you free from the microbes with which you are working. Frequent washing of hands is a necessary part of sterile technique. This is especially important if you are working with unknown microbes. Because some of the microbes found in our environment may be pathogenic (disease-causing), you must take extra care. *Always consult with your teacher or advisor about proper handling and disposal of all microbes purchased or cultured in these projects.*

SOURCES OF MICROBES

Because microbes are too small to handle as individual cells, you will deal with cultures that are simply large numbers of the microbes growing in clumps on some kind of media (discussed below). The microbe cultures are grown or purchased in vials with screw caps or petri dishes with lids, that contain the culture media. If a culture is purchased, it might come in vials containing a broth or a solid nutrient agar. These vials with agar are called *agar slants.* They might come growing in a petri dish that contains a nutrient agar that acts as the culture media. These petri dishes are often simply called *plates.* (See Fig. 1-3.)

CULTURE MEDIA

When you purchase a culture, whether it be in vials or plates, the microbes are growing on some form of culture (growth) media. When you grow your own cultures, they too must be grown on a culture media. This culture media supplies all the nutrients required by the microbes to survive and gives them a place to reproduce.

Most of these experiments require you to grow microbes that are not contaminated with other types or sources of microbes. This means you must use a *sterile culture media.* This is a food for the microbes that doesn't have any microbes on it before you add (inoculate) the microbes to be studied. Sterile culture media can be purchased from a biological supply house. (See the list of these scientific supply houses in the back of this book.)

There are probably as many recipes for microorganism foods (culture media) as there are microbiologists. In this book, you'll find a few standard recipes for culture media that many microbes can grow on. *Nutrient agar* and *tryptic soy agar* are two common examples. Some projects require special media. The requirements are listed in the Materials section of each project. You can purchase

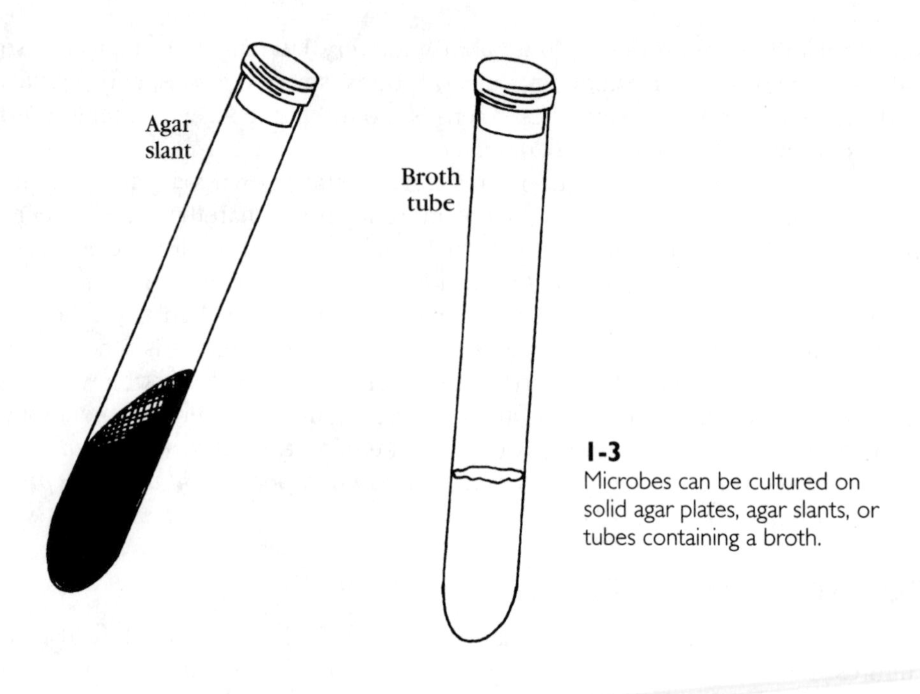

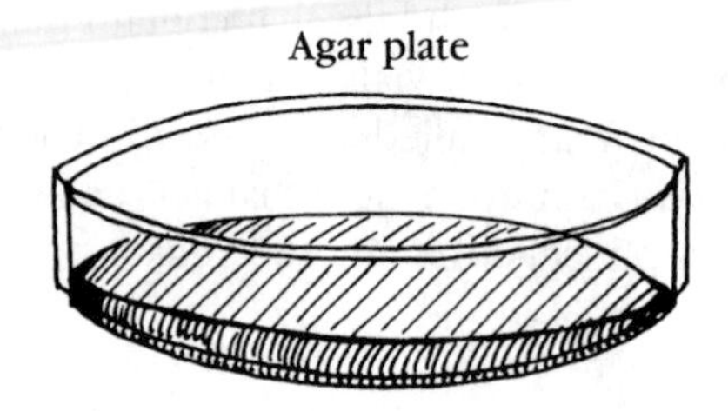

1-3
Microbes can be cultured on solid agar plates, agar slants, or tubes containing a broth.

all of these media from scientific supply houses that are listed at the back of this book. If your school has a pressure-cooker, it can be used to prepare the media. Instead of purchasing a package of media that must be mixed and prepared, it is easier (but more expensive) to buy presterilized media that comes in a sterile vial or in a sterile petri dish plate, ready for use.

HOW TO STERILIZE GLASSWARE

If the experiment calls for sterile glassware, such as mason jars, start with clean glassware. Use soapy water to wash glassware, then rinse with tap water several times to remove the soapy residue. Sterilize the glassware by filling them one-half full with water and placing them in a 4-quart pot that has 2 to 3 inches of water. (The water line of the mason jars must be lower than the water line in the large pot.) Place the mason jar lids into the large pot and boil everything for 30 minutes.

When complete, empty the mason jars of the hot water by picking them up with sterile tongs and dumping out the water. However, place these empty jars *back* into the hot water until ready to use. Also, remove the lids from the pot and put them to the side on a sterile surface to dry. The empty glassware will remain sterile in the warm water for about 30 minutes. This is because the rising hot air from the water in the pot prevents any microbes from settling on the glassware.

This sterilization procedure will not kill bacterial spores. These must be killed by an *autoclave* (a machine that produces high-pressured steam). However, for most of the projects in this book, the glassware can be sterilized by boiling as described above.

Another way to sterilize glassware is to seal the jars with aluminum foil and bake in an oven for 20 to 30 minutes. This technique can be used for petri dishes and pipettes as well. Always remember that a sterile jar can become nonsterile when exposed to air. If the jar is not hot, and is open to the air, it can become contaminated. The longer a container is open, the more likely it will become contaminated. Whenever you conduct microbiology experiments, consider where and how the sterile glassware and tools could get contaminated and do your best to prevent this from happening because it will ruin your project.

HOW TO USE AN INOCULATING LOOP AND SPREADING ROD

Transferring microbes from a vial or dish to a sterile culture media is called *inoculation.* To transfer microbes, you must transfer them on a sterile utensil from their source container to their new container. An *inoculating loop* is generally used for this purpose (see Fig. 1-4). The loop is simply a metal (usually platinum) or plastic loop, connected to a long handle. It is easiest to use plastic, presterilized inoculating loops that are available from scientific supply houses. They are designed to be used once, then thrown away.

If you are using a platinum inoculating loop, it must be *flamed* in a Bunsen burner to become sterilized before each use. *Always work with your teacher or advisor when a project calls for the use of flames.* The loop is placed in the flame until it becomes red-hot, which means it is now sterile (free of microbes). Let the loop cool for 10 seconds or so. (If the loop is too hot when it enters the culture it will kill many or all of the cells and the transfer will not be successful.) Do not

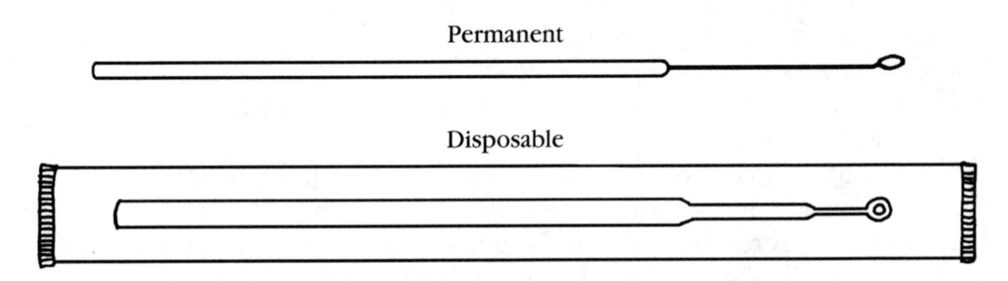

1-4 Presterilized, disposable inoculating loops are available or long lasting loops can be sterilized before each use.

wave the loop around in the air when it is cooling or lay it on a surface before use. This would recontaminate it.

Once the loop cools, you are ready to transfer a sample of the culture (called *inoculum*) from the source to the vial or plate containing sterile culture media. Dip the loop into the vial or plate containing the culture to be transferred. A sample of the culture will remain on the loop. To transfer it to a vial, simply stick the loop carrying the sample into the sterile vial. To transfer it to a plate containing sterile culture media, such as agar, gently move the loop over the surface of the media. Don't dig the loop into the surface of the media in the plate (see Fig. 1-5).

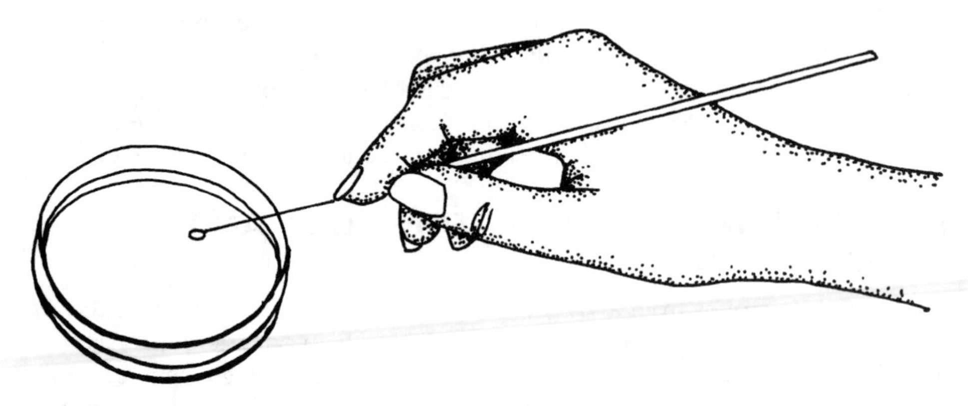

1-5 Gently touch the inoculating loop to the plate.

Once you have transferred (inoculated) the culture, you must discard the loop if it is disposable or remember to reflame the loop if it is metal. If you put the loop down on the counter after the transfer, you contaminate the work surface and you spread microbes around the room. So put the loop back into the flame after it has been used before laying it on the counter.

A spreading rod is required for some of the projects in this book. This is simply a metal, glass, or plastic rod with a bend in it. The longer, top portion is the handle, while the bottom portion, beneath the bend, is used to spread a liquid culture (inoculum) over the surface of the media found in a plate (see Fig. 1-6).

Sterilizing a spreading rod must be performed by your teacher or advisor. The spreading rod must be sterile or the culture will become contaminated. This is usually done by dipping the spreading rod in alcohol and lighting the rod to burn off any contaminants.

To sterilize a spreading rod, your teacher should dip the bottom portion in alcohol and ignite the alcohol on the loop with the flame until all the alcohol is burned off. The spreading rod must be pointed downward while the alcohol is burning off so it doesn't drip on the person holding it. Keep the rod away from the flame and away from the container of alcohol while it's burning. Also keep the container of alcohol away from the open flame.

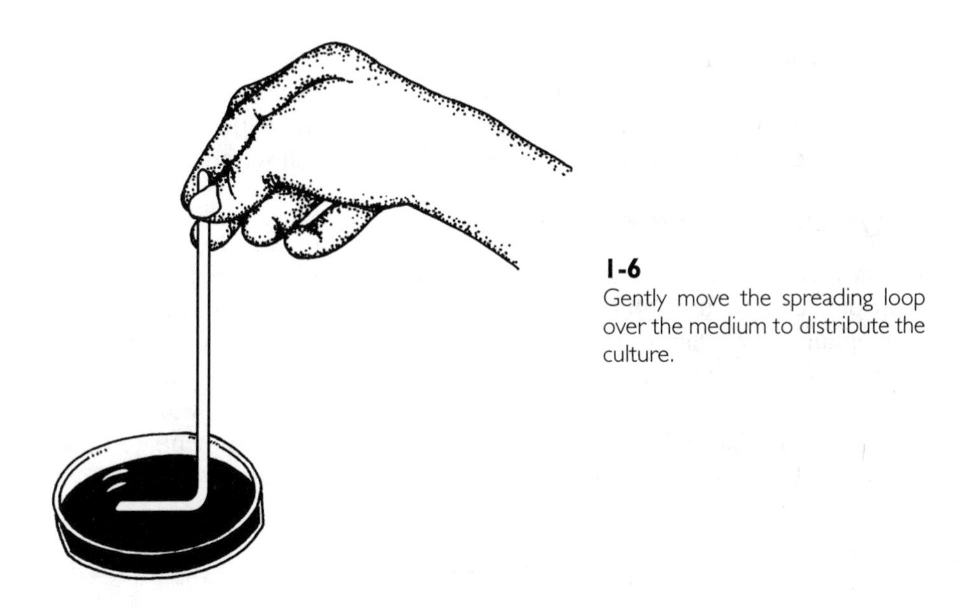

1-6
Gently move the spreading loop over the medium to distribute the culture.

Use the sterile spreading rod to gently spread the culture equally over the surface of the plate. Do not set the spreading rod down on the work area, because it might become contaminated.

USING A PIPETTE

A *pipette* is a long, thin tube with markings that show the volume in the tube and a bulb at one end used to bring fluids into the tube by suction (see Fig. 1-7). It is used to transfer liquids and looks very much like an eyedropper. When a pipette is used to transfer a liquid culture of microbes it must be sterile or the resulting culture will be contaminated.

Try to use sterile, disposable pipettes that can be purchased from a scientific supply house. They are not expensive and can save a lot of lost time or failed experiments because of contaminated pipettes. Sterilization of pipettes is difficult.

You should never pipette by mouth. Instead, use the pipette bulb, which works just like an eyedropper bulb to bring the fluid into the pipette. Some pipettes have a *pipette man* (a manual pump) to suck liquids up into the tube.

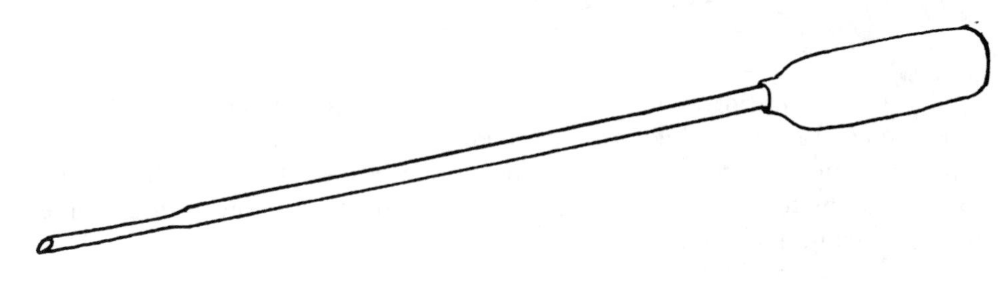

1-7 Presterilized, disposable pipettes are easy to use.

Pipetting by mouth is very dangerous because some of the fluid can get into your mouth. Furthermore, it is too easy to contaminate a culture when pipetting by mouth. You might need a little practice when working with a pipette bulb or pipette to accurately transfer small volumes, but it is well worth your time.

INCUBATING CULTURES

Incubating a culture means keeping the plates or vials that contain cultures at the proper temperature. This is usually done by placing them in an incubator that maintains a constant temperature (see Fig. 1-8). Your school might have an incubator. If not, almost all of the microorganisms used in the experiments in this book can be incubated by simply placing the cultures in a warm area in a room or under a lamp. Each project gives the temperature to incubate the cultures and states how to incubate them without an incubator.

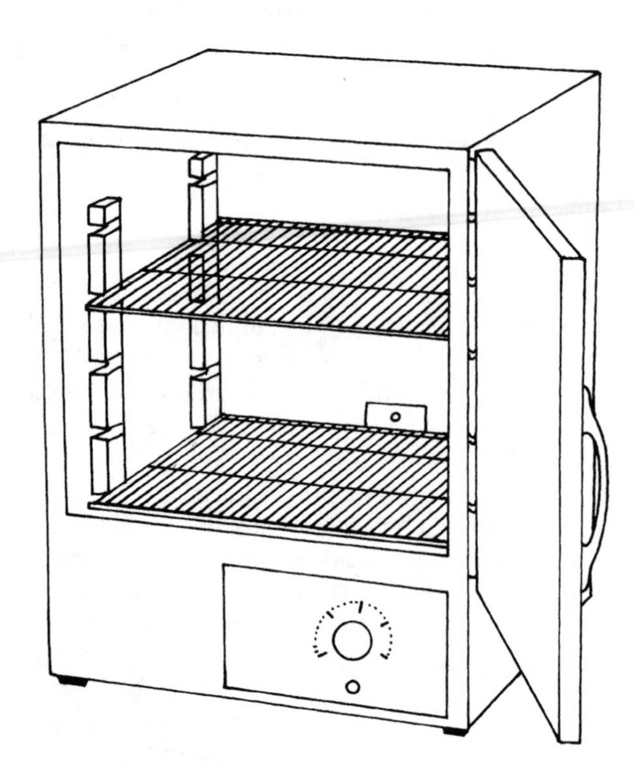

1-8
An incubator is the best way to maintain the correct temperature for your cultures, but not the only way.

If you are not using an incubator, be sure to test the temperature of the area to be used before beginning any experiment. Do this by placing a thermometer in the area and reading it throughout the course of the day and night. It is important that the temperature of the area does not fluctuate more than 2 degrees from the required temperature.

COUNTING MICROBE CELLS

Some of the projects in this book require you to count the number of cells in a certain volume of liquid. The easiest and most accurate way to do this is with a *hemocytometer.* A hemocytometer is a special microscope slide that contains a known volume, under which is a tiny, tiny grid. Put a drop of the culture you need to count in the hemocytometer, cover with its coverslip, and then count all the cells in the known area to get an estimate of the total number of cells in a known volume. A hemocytometer is expensive and might not be available at all schools.

If a hemocytometer isn't available, you can estimate the number of some types of cells (e.g., algae, fungi, protists) by placing a small piece of grid paper (1mm by 1mm) beneath the slide. Using low power, you can then count of the number of microbes in that area. By knowing the volume of culture placed beneath the coverslip (e.g., .1 ml from a pipette) and the area under the entire coverslip (usually 22mm×22mm) you can determine a rough estimate of the number of microbes present in the entire .1 ml sample.

2

Getting started

Selecting a project & using the scientific method

The best way to select a project is find out what interests you about microbes, if you don't already know. Have you ever wondered what was actually in that green murky water in your fish tank or in a nearby pond? Did you ever wonder what hidden worlds can be discovered in a handful of dirt? How about what *germs* you are spreading when you sneeze, or what's in your dog's sloppy drool?

Stop and let your mind wander for a while. What comes to mind? It could be anything, anywhere, anybody! Once you've opened your mind and let your imagination run wild, look through the Contents in this book for more specific topics to research. Read through the Overview sections (the first few paragraphs) of each project for additional information. Select a project that you are not only interested in, but truly enthusiastic about.

SCIENCE FAIRS

Science fairs give you the opportunity to not only learn about a topic, but to participate in the discovery process. Although you probably won't discover something previously unknown to humankind, you will perform the same processes by which discoveries are made. Most science fairs have formal guidelines or rules. For example, there might be a limit to the amount of money spent on a project or the use of live organisms. There might be regulations on the use of certain microbes. Be sure to review these guidelines and check that the experiment poses no problems.

THE SCIENTIFIC METHOD

The scientific method is the basis for all experimentation. It simply, yet clearly, defines what scientific research is all about. The scientific method can be divided into five steps.

Purpose

What question do you want to answer or problem would you like to solve, for example: "How far can microbes be dispersed when you cough?" or "Can ants track microbes all over your house?". The overview section (which appears as the first few paragraphs of each project) offers thought-provoking questions and problems.

Hypothesis

The *hypothesis* is an educated guess, based on preliminary research, which answers the question posed in the purpose. You might hypothesize that microbes can travel 30 feet when you cough or that ants can track microbes around your house.

Experimentation

The *experiment* determines whether or not the hypothesis was correct. Even if the hypothesis wasn't correct, a well-designed experiment would help determine why it wasn't correct.

There are two major parts to the experiment. The first is designing and setting up the experiment. For example, how must the experiment be set up and what procedures must be followed to test the hypothesis? What materials will be needed? What cultures, if any, are needed? What step-by-step procedures must be followed during the experiment? What observations and data must be made and collected while the experiment is running? Once these questions have been answered, the actual experiment can be performed.

The second part is performing the experiment, making observations, and collecting data. The results must be *documented* (written down) for study and analysis. The more details, the better.

The Materials section of each project lists all the materials needed for each experiment, and the Procedures section explains how each experiment is to be performed. Suggestions are given on what observations should be made and what data should be collected.

Research

The *research* part of the project should begin before you start the actual step-by-step experiment and continue after you have collected the results. Read as much as you can about the topic you are studying. Use any and all sources available to you and try to be the expert on the subject. Once the experiment is completed, analyze the results and see how what you have learned compares with what is already known about the subject.

In addition to researching the primary topic, the Going Further section gives you ideas on related topics to research.

Conclusion

Once you have collected and analyzed the data and researched the subject, you can draw your own conclusions. Creating tables, charts, or graphs will help you analyze the data and draw conclusions from it.

The conclusions should be based upon your original hypothesis. Was it correct or incorrect? Even if it was incorrect, what did you learn from the experiment? What new hypothesis can you create and test? Something is always learned while performing an experiment, even if it's how *not* to perform the experiment the next time.

Part two

Microbes of humans & animals

Microbes live with us, whether we are aware of them or not. They are in the air we breathe, in the food we eat, and in the water we drink. They live on our skin, in our bodies, as well as on and in our pets. Luckily, most of these microbes never cause us any health problems.

The first two projects in this section investigate the microbes on our hands and body. The next few projects explore microbes on our pets. Advances in microbiology and the treatment of infectious diseases have also helped our pets, including dogs and cats. They are surrounded by microbes just as we are. Fish too, live in a world filled with microbes. Microbes often attach themselves to the gills of a fish. As long as the numbers do not become too large, this is not a problem for the fish. But in large enough numbers, a fish can quickly die from microbes.

The other projects in this section investigate microbes that live with us in our cities. We don't generally think of cities as being filled with living organisms, other than humans, but they are. A city has plenty of places for microbes to live. Air transports microbes from place to place. City water is inhabited with microbes, although the number should be low. City soils may be poor in nutrients, but usually are also inhabited with microbes. Microbes can be found all around a city: fountains, garbage cans, and between the cracks of the sidewalks. Even in an urban setting, there is a teeming microbial world.

3

Scrub up

What method of washing your hands best reduces the number of microbes?

We carry microbes with us every day. Our skin is our armor against microbes that prevents them from entering our bodies. Our skin doesn't kill the microbes—it just blocks their entrance. Whenever we touch things, we are picking up microbes. This is why hospital staffs constantly clean their hands, so they don't pass *pathogens* (germs) from one patient to another. This is also why restaurant employees should always clean their hands, so they don't pass microbes from themselves to the food. And this is why mothers tell their children to wash their hands. How clean do you think your hands are? What is the best way to wash your hands to eliminate microbes?

MATERIALS

- Eight sterile nutrient agar petri dishes (available from scientific supply house)
- Marker that can write on plastic
- Hot and cold tap water
- Paper towels
- Bar of bath soap
- Bar of bactericidal soap (available at a pharmacy or drug store)
- 1-inch adhesive tape
- Incubator (or warm area in room under lamp)
- Thermometer
- Magnifying glass or stereoscope

PROCEDURES

Don't wash your hands for at least 4 hours. Prepare or purchase sterile nutrient agar plates so they are ready to be inoculated. Open the cover of one plate slightly with your right hand and touch the finger tips of your left hand to the surface of a plate for 1 second (see Fig. 3-1). Close the plate and label it on the bottom of the plate, "No Wash, Left." Repeat this procedure with your right hand and label it on the bottom of the plate, "No Wash, Right."

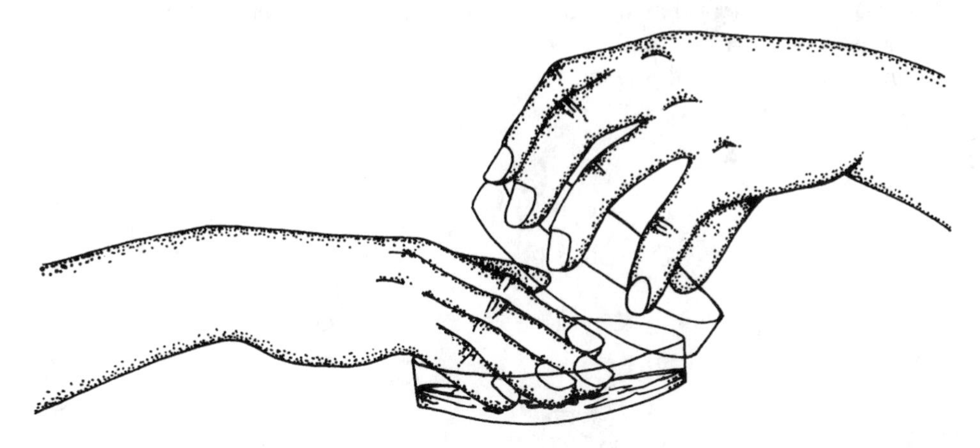

3-1 While holding up the lid with one hand, gently touch your fingertips to the agar with the other hand.

Now rinse your hands under cool running water for 10 seconds and damp them dry with a towel. Repeat the same process as before, this time with two new plates. Close these plates and label one "rinse, left" and the other "rinse, right."

Next, use a bar of bath soap and lukewarm water to clean your hands for 30 seconds and repeat the process once again. Label these plates "Soap, Left" and "Soap, Right." Finally, wash your hands again, with hot water and bactericidal soap for 2 minutes. Repeat the process one final time for both hands and label the plates "Bact. Soap, Left" and "Bact. Soap, Right."

Seal the plates by taping their edges closed with adhesive tape. Place in an incubator, if available, at 37 degrees C or keep the plates in a warm area in the room. You might need to place them under a lamp. Note the temperature of the area and check it periodically to be sure there is no more than 2 degrees fluctuation. Observe the plates each day and make notes about microbial growth on the plates. Each day, note the shape of the colonies, their color, position on the plate, and the size and number of colonies. Observe the colonies through a magnifying glass or a stereoscope if it is available.

CONCLUSIONS

How did the colonies differ between each group? Was there a noticeable difference in the number of colonies? How about different kinds of colonies? Does washing your hands really make a difference in controlling the numbers and types of microbes on your hands? Does using warm water and certain soaps make a difference?

GOING FURTHER

Design an experiment that would determine how soon after washing your skin microbes return. How do different activities (e.g., playing outside, cooking, reading indoors) affect the results?

4

Out of sight —out of mind

Which parts of your body do the most microbes call home?

Even though our skin protects us from microbes entering our bodies, some microbes live on our skin all the time. When you have skin infections, it is because microbes have gotten into a hair follicle or a tiny cut and have begun to multiply. Are microbes present everywhere on your body in equal amounts or do different areas of your skin contain different numbers or types of microbes? How does bathing affect these microbes?

MATERIALS

- At least 10 sterile swabs (available from scientific supply house, pharmacy, or drug store)
- At least 10 sterile nutrient agar petri dishes (available from scientific supply house)
- Marker that can write on plastic
- Bathtub (or shower)
- Bath soap
- Bath towel
- 1-inch adhesive tape
- Incubator (or warm area in room under lamp)
- Thermometer
- Magnifying glass or stereoscope

PROCEDURES

Start this experiment at least 6 hours after your last bath or shower. Take a sterile swab and rub it over 1 inch of skin on the back of your hand (see Fig. 4-1). Immediately take the swab and gently roll it over the entire surface of a nutrient

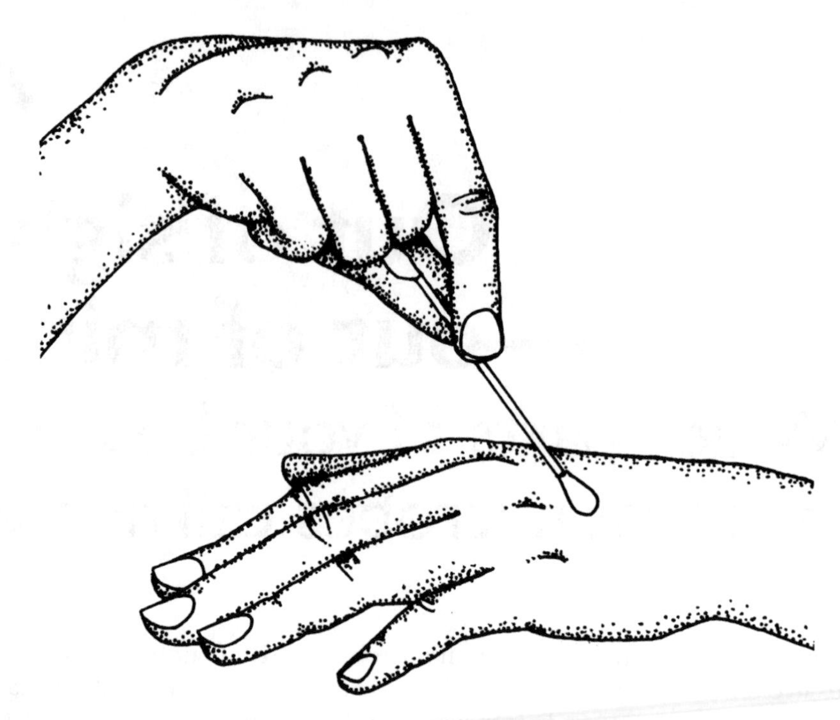

4-1 Rub the cotton swab over the skin on the back of your other hand.

agar plate and close plate. Label the plate "B/B—Back of Hand." (The b/b is for "before bath.") Repeat, taking swabs from various parts of your body (e.g., forehead, underarm, back of knee, sole of foot). Be sure you only take a 1-inch swab from each body section. Label each plate on the bottom with the "b/b" and the name of the body section where the swab was taken. Now take a bath or shower the way you normally do, using the same soap(s) as usual.

After you have dried off, repeat the process by taking swabs in the exact same regions of your body as you did before the bath or shower. Label these plates "A/B" (for "after bath"), followed by the body region. Now close all the plates by taping their edges with adhesive tape. Place in an incubator at 37 degrees C (if available) or keep the plates in a warm spot in the room. You might need to place it under a lamp. Note the temperature of the area and check it periodically to be sure there is no more than 2 degrees of fluctuation.

Observe the plates each day and make notes about microbial growth on the plates. Each day, note the shape of the colonies, their color, position on the plate, and the size and number of colonies. Observe the colonies under a magnifying glass or a stereoscope, if available.

CONCLUSIONS

How many similar types and different types of colonies were produced from each portion of your body before bathing? Did certain parts of your body contain more microbes? What happened after bathing? Did the numbers and/or types of colonies change? Did they all change in the same way, or did some areas show different amounts of change?

GOING FURTHER

Compare the difference between a shower and a bath. Does one get you cleaner than the other? Compare the difference between microbes found on your skin in different types of weather conditions (e.g., on a hot, dry day versus a hot, humid day). Does weather affect the number of skin microbes?

5

Spit & drool

Microbes in the saliva of people & dogs

Saliva has many functions. It is primarily used to begin the digestion of the food we eat. In addition, saliva moistens the mouth and the tongue, and aids in tasting and talking. It also washes microbes off teeth and gums, and acts as an antimicrobial, meaning it reduces the number of microbes in our mouths. Do different animals have different numbers or types of bacteria living in their saliva?

MATERIALS

- Two sterile teacups (follow procedures in chapter 1 "An Introduction to Microbiology" on how to sterilize glassware)
- Marker
- Four sterile tryptic soy agar plates (available from scientific supply house)
- Four sterile blood agar plates (available from scientific supply house)
- Two sterile 1 ml pipettes graduated with 0.1 ml (follow procedures in chapter 1 "An Introduction to Microbiology" on how to sterilize pipettes)
- Your dog or your neighbor's dog (be sure the dog is in good health before proceeding)
- Sterile spreading rod (follow procedures in chapter 1 "An Introduction to Microbiology" on how to sterilize spreading rods)
- Incubator (or warm area in room under lamp)
- Jar (with an opening wide enough to fit a petri dish)
- Small candle in glass votive

PROCEDURES

Label one teacup "human" and the other "dog" (see Fig. 5-1). Collect saliva from yourself by spitting into a small, clean teacup. Using the sterile pipette, collect 0.2 ml of saliva. Collect the same amount from a dog. Dogs will usually salivate when they see food, so show the dog some food, but don't let it eat it until you

5-1 Label one of the cups "dog" and the other "human."

have collected a couple drops of saliva. You may want a friend to help you hold the dog while collecting the saliva.

Label two of each type of agar plates "Human" and the other two "Dog." Using a sterile pipette, put 0.1 ml of your saliva on each "Human" plate. Spread the saliva evenly over the surface of the plate using a sterile spreading rod. Put 0.1 ml of the dog saliva on each of the "Dog" plates using the same technique.

Incubate one of each of the plates at 37 degrees C. Place in an incubator at 37 degrees C (if available) or keep the plates in a warm spot in the room. You might need to place it under a lamp. Note the temperature of the area and check it periodically to be sure there is no more than 2 degrees of fluctuation.

The other two plates must be placed into a CO_2 jar before being incubated (see Fig. 5-1). Because this jar requires a flame, have your teacher or advisor do this part for you. To make a CO_2 jar, you need a large jar that has a mouth wide enough to fit a petri dish through it. The jar must have a tight, screw-on lid. Place the plates to be incubated at the bottom of the jar and then have your teacher or advisor put a lighted candle, which is in a glass container (glass votive), on top of the plates as illustrated in Fig. 5-2. Close the lid to the large jar. The candle will use most of the oxygen in the jar before it goes out. This creates a slightly *anaerobic environment* (low oxygen environment). Incubate the entire CO_2 jar with the enclosed plates in the same manner as the other two plates.

Incubate all the plates for 3 to 4 days. *Do not open these plates.* Then count the numbers and types of colonies on each plate. Place a mark on the glass top above each colony as it is counted. This will prevent you from counting the same colony twice.

CONCLUSIONS

Do you get different numbers or types of colonies in the plate containing human saliva compared to the dog? Is the number of colonies different on the different media types (i.e., blood versus soy agar)? Does a limited amount of oxygen affect the numbers or types of colonies? What do you conclude about the microbes in your saliva and the dog's saliva?

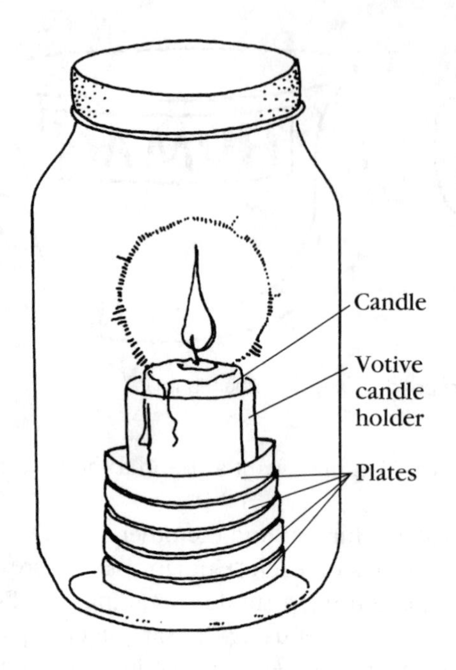

5-2
The flame uses up most of the oxygen in the CO_2 jar, creating an anaerobic environment for the microbes in the petri dishes.

GOING FURTHER

Run a similar experiment, but collect the dog's saliva at different times—before and at intervals after eating or after the animal has been out and about for a while. Do you get the same results?

6

Scaly world

What kinds of microbes are found on dead goldfish?

Microbes thrive in a moist world. People living in humid climates have a lot more skin infections than people in dry climates. Homes with moist air will have more mold growth than those with dry air.

A fish's home is constantly wet. Fish must combat microbes in their environment all the time. If fish lose this battle, they become diseased and die. What types of microbes attack fish? What kinds of microbes are found on a dead goldfish? Are the same types of microbes found on different parts of the fish?

MATERIALS

- Rubber gloves
- Three dead goldfish (available from a local pet store)
- Bottom of one petri dish or similar type of dish or saucer
- Forceps
- Microscope slides
- Water
- Eyedropper
- Coverslips
- Drawing paper and pencil
- Book to identify aquatic microbes (protozoans and fungus especially)

PROCEDURES

While wearing rubber gloves, place the dead fish in a petri dish or similar saucer. Use the forceps to remove a scale from the dead fish (see Fig. 6-1). The best place to remove the scale would be around the gills. Place the scale on a microscope slide. Put a drop of water over the scale with the eyedropper. Cover the drop with the coverslip. Look at the slide under the microscope, first at 100×, and then at 400×.

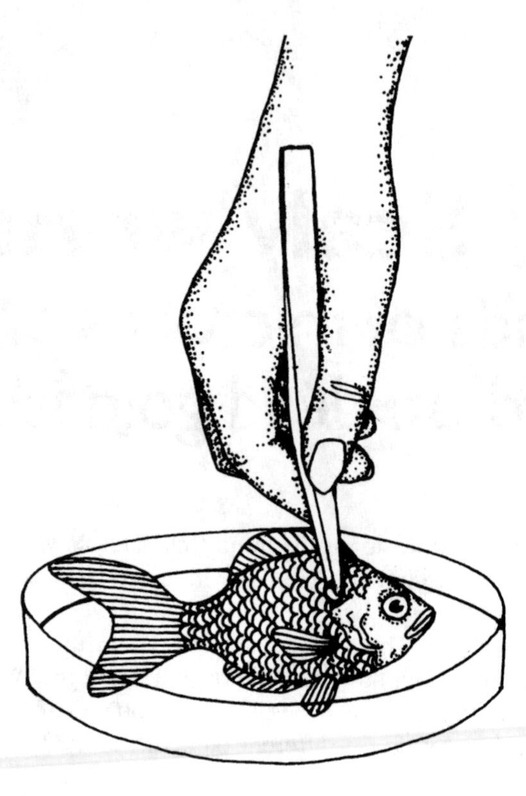

6-1
Use a forceps to pick a scale off of the gills of the dead goldfish.

Draw the shapes of any microbes you see. Repeat this with scales from other parts of the fish's body and with scales from other dead fish. Use an bacteria and fungus identification book to help you identify the microbes.

CONCLUSIONS

What kind of microbes do you find on the fish? Did you find different types of microbes on scales from different parts of the fish? Did you find different types of microbes on different fish? Why were these microbes on the dead fish? Would they be found on a live fish?

GOING FURTHER

Collect dead fish from different aquaria and from natural habitats and look for different types of microbes. Collect scales from a fish that has *ick*, short for *Ichthyophthirius multifilius*—a common fish parasite that appears as small salt-like white granules on the fish's body and fins—or some other fish disease. What did you find on them? Look for microbes on the scales of a live fish. Check with your teacher for proper handling procedures. You should not have to harm the fish to look at them under the scope for a short while.

7

The four seasons

Seasonal changes in microbe populations in your backyard

Seasonal changes in living creatures are often very easy to see. For example, trees lose their leaves in the fall, grass gets greener in the spring, and flowers bloom in the summer. Seasonal changes in microbes, however, aren't so easy to see. Are there seasonal changes in the numbers and types of microbes found in the air? Can you find differences in microbe populations at different times of the year?

MATERIALS

- 36 sterile nutrient agar plates (available from scientific supply house)
- 36 sterile Sabouraud agar plates (a special type of culture media available from scientific supply house)
- 1-inch adhesive tape
- Incubator (or warm part of the room with lamp)
- Drawing paper
- Pencil
- Magnifying glass or stereoscope

PROCEDURES

This experiment will take 3 months. Follow the subsequent procedure once each week for the 3-month period. It must be done at the same time of the day throughout this period. Take three plates of nutrient agar and three plates of Sabouraud agar into your backyard. Open one plate at a time and pass it through the air by waving your arm back and forth two times (see Fig. 7-1). Then close the plate. Do this for all six plates, then seal them with adhesive tape. Label the bottom of the plates with the date and type of media (i.e., nutrient or Sabouraud). Incubate the plates for 2 days at 37 degrees C or in a consistently warm area of the house, such as under a lamp. (Be sure the temperature does not fluctuate more than 2 degrees.)

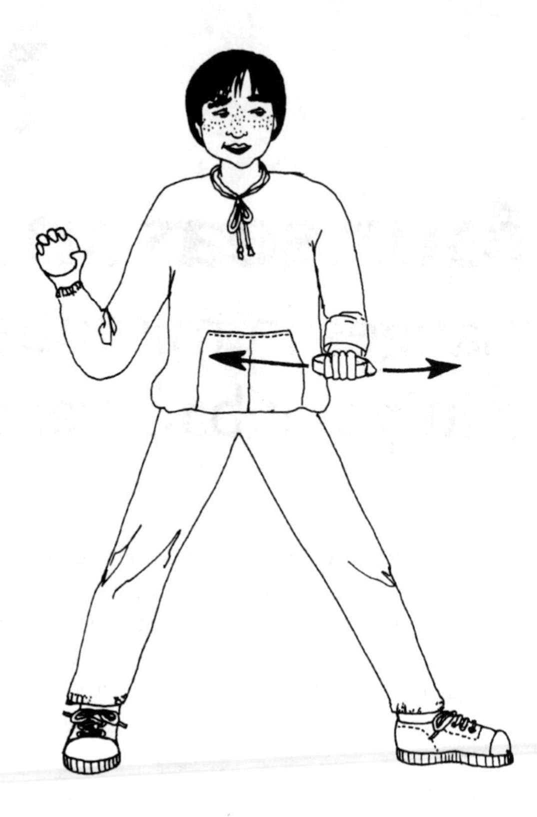

7-1
Hold the cover of the petri dish in one hand, while waving the bottom half back and forth.

Note the shapes and colors of the colonies on each plate. Draw the different types of colonies. You need a detailed written description of their forms, because you will be comparing colony types over a 3-month period. Observe the colonies under a magnifying glass or a stereoscope, if available.

CONCLUSIONS

First, look for changes in the types of colonies over the 3-month period on one type of media. Did the numbers and types of microbes differ throughout this period? Plot a graph of the results for each type of media.

Then, compare the differences between the two types of media. Do you see the same types of colonies on both of the plates incubated in the same week? Are there differences between the two groups over the period? Do the different media give rise to different types of microbes? If so, do the differences appear all or just some of the time? What do you conclude about seasonal changes with airborne microbes?

GOING FURTHER

Repeat this experiment in a different season or take samples at different times of the day. Do you get the same results?

8

Life in the AC

Microbes on dirty air-conditioning filters

Urban air is often dirty. Exhaust fumes can irritate eyes and soot can collect on windows. Air conditioners have filters to collect some of this dirt before it gets into the cooling system. Does an air-conditioning filter also collect microbes? What types of microbes can be found in an air-conditioning filter?

MATERIALS

- Marker to write on plastic
- Two sterile Sabouraud agar plates (a special type of culture media available from scientific supply house)
- Two sterile nutrient agar plates (available from scientific supply house)
- Scissors
- Sterile forceps (follow procedures in chapter 1 "An Introduction to Microbiology" on how to sterilize utensils)
- Used air-conditioning filter (take a well-used filter out of an air conditioner)
- New air-conditioning filter (purchase a new, unused filter for the same air conditioner)
- 1-inch adhesive tape
- Incubator (or warm area in room under lamp)
- Paper
- Pencil

PROCEDURES

Label the bottom of the Sabouraud plates: "S Used" and "S New." Label the bottom of the nutrient agar plates: "N Used" and "N New." While holding the used air-conditioning filter with sterilized forceps, cut two 1-inch-square pieces from a part of the filter that looks dirty. Place one of these filter pieces in the center of a nutrient agar plate and the other in the center of a Sabouraud agar plate (see

Fig. 8-1). Close the plates and seal with adhesive tape. Repeat this same procedure with a similar piece of new air conditioner filter. Incubate all the plates at 37 degrees C or in a consistently warm area of the house, such as under a lamp. The temperature cannot fluctuate more than 2 degrees.

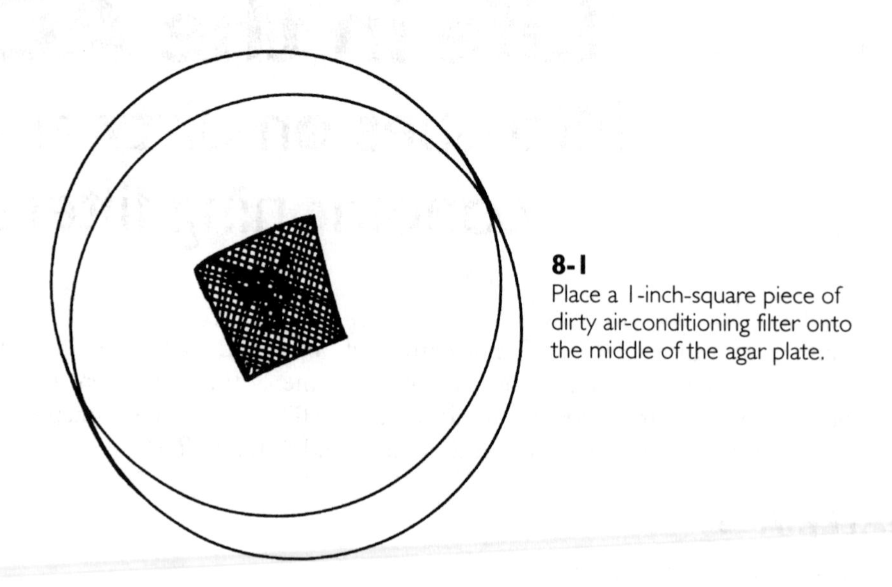

8-1
Place a 1-inch-square piece of dirty air-conditioning filter onto the middle of the agar plate.

CONCLUSIONS

Look at the plates after 3 to 4 days. Is there any growth? Compare the numbers and types of colonies found. Draw pictures of each plate. Are there differences between the new and old filters? What do you conclude about the ability of the air-conditioning filter to collect microbes? Does it work? Can you draw any conclusions about the presence of microbes in the air within your house?

GOING FURTHER

Repeat this experiment with an air conditioner that has been shut off for more than 1 month and one that has been off for more than 3 months. Can microbes survive for a long time in an air-conditioning filter?

9

Exhaustive study

The effect of air pollution on lichens

Lichens are truly unusual types of organisms. They are a fungus and an alga, living together in a *symbiotic relationship*—neither could survive without the other. The fungi provides protection for the algae, keeps them from drying out, and gives the algae a secure home. The algae provides food for the fungi. It is a nice arrangement for both.

Lichens can be easily classified by their shape (see Fig. 9-1). A lichen that forms a crusty, flat layer over a large rock or a tree is called a "crustose" lichen. A lichen that forms masses in upright stalks is called a "fructicose" lichen. "Foliose" lichen have flat areas and are leaf-like with curled up edges. Squamulose are leaf-like.

Lichens get many of the chemicals needed for life from the air like most organisms. Because lichens depend on the air to survive, are they affected by air pollution?

MATERIALS

- Notebook
- Woods adjacent to a highway (a highway that has been recently closed is best)

PROCEDURES

With a notebook in hand, find a section of woods that has an edge along a highway. (Warning: Don't get too close to the road. Get approval from your teacher or other adult advisor about the location and have an adult with you during this part of the project.) Look for lichens in the woods close to the shoulder of the highway. Look for lichens on rocks and trees. Note the number and the type of lichens (as described above) you find on each tree or rock. Do this until you have counted the lichens on 20 trees or rocks.

Now, walk deeper into the woods, about 100 more feet away from the road, and look for lichens in a line parallel to the road on 20 different trees or rocks.

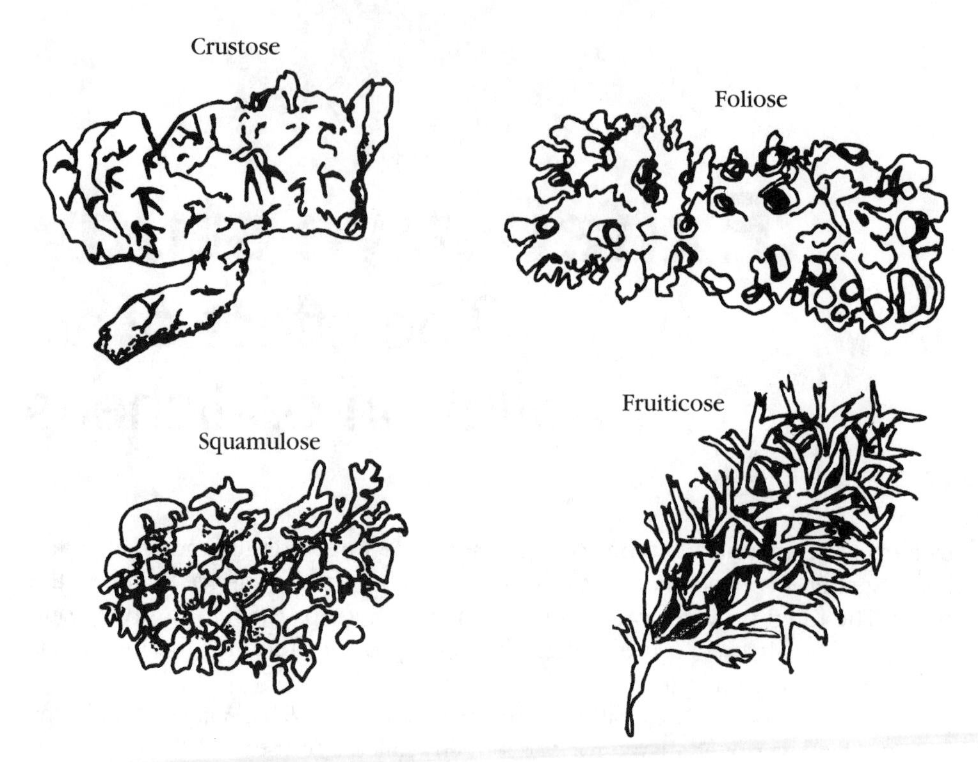

9-1 Lichens can be found in four basic forms.

Again, note the number and type of lichens on each tree or rock. Walk another 100 feet more into the woods and repeat this once again.

CONCLUSIONS

Do the numbers and types of lichens change as you go further from the road? Do the numbers for each of the four types of lichens increase or decrease as you get further away from the road? From this data, do you think all or some types of lichens are affected by car exhaust and air pollution?

GOING FURTHER

Study the amount of pollutants found on the leaves at varying distances from a road. To do this, take swab samples from leaves and look at them under a scope.

Part three

Microbiology of food

The study of food microbiology is important for two reasons. First, microbes are needed for the production of many of our foods, such as fermented foods. Second, microbes can spoil many of our foods and may cause disease in the process.

Microbe growth must be controlled in our foods to prevent the contaminates from using our foods before we do and also to prevent *toxins* (poisons) that can live in the food. Microbes in foods can be controlled by either reducing the number of microbes that already exist (e.g., through processes such as pasteurization) or by creating conditions that prevent the microbes from growing in the first place (e.g., freezing). The first three projects in this section investigate preserving our foods from microbes.

Diary products can be either spoiled or made delicious by microbes. It all depends on the type of microbe and your choice of dairy product. Soured milk is an example of a dairy product that has been spoiled by microbes. Sour cream, buttermilk, blue cheese, and yogurt are examples of dairy products that are improved by microbes. The final project in this section investigates how yogurt is created with microbes.

10

The spice of life

Dehydration, spices, & the preservation of foods

Some microbes cause food to spoil, while other microbes result in food poisoning. Food spoilage results in the loss of vast amounts of food, while food poisoning causes illness or even death. Food spoilage microbes usually change the appearance and taste of food, but food poisoning is not so easy to see or taste.

In an effort to protect people and our domesticated animals from food poisoning and food spoilage, scientists have developed many methods to reduce or prevent these microbes from attacking our food supply. How can we slow or prevent food spoilage? Does dehydration (drying) of foods protect the food against spoilage? Do spices, such as pepper, prevent or delay spoilage from occurring?

MATERIALS

- Bread containing no preservatives (eight slices from the same loaf)
- Plastic bags (e.g., Ziploc sandwich bags)
- Toaster
- Marker (that can write on plastic)
- Tablespoon measurer
- Pepper
- Notebook and pencil
- Magnifying glass or stereoscope

PROCEDURES

Take four fresh slices of the bread and place each in a separate plastic bag. Then seal all the bags. Take the other four slices of bread and toast them to dryness. Place each of these slices in separate plastic bags and seal all these bags. Label two of the bags containing toasted slices and set them aside. Label two of bags containing fresh bread and set them aside. This leaves four bags—two toasted and two fresh.

Add 1 tbsp. of pepper to each of these four bags and reseal all of them. Gently shake the bags so the pepper covers all parts of the bread. Label the two bags with toasted bread and pepper—dry and spiced. Label the other two bags—fresh and spiced. Be sure the bags are well sealed (see Fig. 10-1). Place them on your kitchen counter or in a bread box.

10-1 You will have two sets of bags similar to these.

Each day, over a period of 6 weeks, look at the bags for signs of spoilage, such as bread mold. Take notes about when the bread appears spoiled, and what the spoilage microbes look like. Do not open the bags. Make your observations through the plastic.

CONCLUSIONS

Does food spoilage occur in all the bags? Does it occur at the same time and as rapidly? Which treatment appears to provide the most protection to the food?

GOING FURTHER

Identify the type of microbes found on bread by opening the bags at the conclusion of the experiment and looking at them under a stereoscope or a magnifying glass, if available. Compare to the microbes found on other foods. Use different spices to prevent the food from spoiling. Do some work better than others? Do some not work at all? What is the effect of temperature on food spoilage? What is the effect of moisture?

11

pH & food

The effect of pH on food spoilage

As described in chapter 10 "The Spice of Life," food spoilage is a major problem affecting the world's food supply. There are many ways that people try to control the microbes that are responsible for this spoilage. One such method is to change the pH of the food. Microbes, like all forms of life, have an ideal acidity/alkalinity level (the pH level) at which they grow the most rapidly. If the pH is changed, by the addition of an acid (e.g., vinegar) or a base (e.g., baking soda), then food spoilage may be slowed or even stopped.

Different foods have different natural levels of acidity. For example, beets are alkaline, while tomatoes are acidic. Can you find differences in the types of microbes that grow in these foods? How does this affect their rate of spoilage?

MATERIALS

- Two fresh beets, sliced
- Two fresh tomatoes, sliced
- Marker (that can write on glass)
- Four sterile, pint-sized mason jars with lids (follow procedures in chapter 1 "An Introduction to Microbiology" on how to sterilize utensils)
- Four spoons
- Pieces of pH paper
- 16 sterile nutrient agar plates (available for scientific supply house)
- Inoculating loop
- Bunsen burner (to be used by your teacher or advisor)
- 1-inch adhesive tape
- Incubator (or warm area of room under lamp)
- Notebook and pencil

PROCEDURES

Label two of the jars "Beets," and the other two "Tomatoes." Place a sliced beet into each of the "Beet" jars and a sliced tomato into each of the "Tomato" jars.

Mash the contents slightly with a sterile spoon. If the fresh beet is too dry add a very small amount of sterile water.

Once all the jars are ready, measure the pH in each jar using pH paper. Then, cover the jars and place them in a warm area in the house. Each day, look for microbial growth in the jars and take the pH of the contents. Note the odor of each jar, and if and when it begins to smell as if it had spoiled.

When the first signs of spoilage occurs in any of the jars, prepare your agar plates for use. Label four nutrient agar plates so there is one plate for each jar per day for the next few days. The plate should be labeled with the matching jar (e.g., "Fresh Beet" and with the day's date) (see Fig. 11-1). Dip a sterile inoculating loop into one of the jars and inoculate the properly labeled nutrient agar plate by moving the loop over surface of the agar, in a straight line.

11-1 The label on the agar plate should contain the date that it was inoculated.

After doing this for one jar, use another disposable sterile loop or have your teacher or advisor resterilize the loop in a Bunsen burner flame and repeat the procedure with the other jars. Seal all the plates. Place in an incubator or under a lamp, to keep their temperature at approximately 30 degrees C. Don't let the temperature fluctuate more than 2 degrees. After 2–3 days, make notes on the types and shapes of colonies on each plate. Take these samples for 4 days following the first sign of spoilage in any of the jars.

CONCLUSIONS

Which begins to spoil first? Are there any differences in spoilage between the jars? How does the pH relate to the spoilage? Does the pH appear to affect when a food will begin to spoil? How do the microbes differ over time from within each jar? Do the numbers of microbes increase as the food begins to spoil? What do you conclude about the effect of pH on microbial growth?

GOING FURTHER

Repeat this experiment with different types of vegetables. Do the microbes appear to be different? How does the original pH of the food affect the outcome?

12

Moldy forms

Mold on breads & cheeses

Molds can be found on food in our refrigerators. Breads and cheeses are two foods that commonly become *moldy* (inhabited by a fungus). Do the molds found on these two different types of foods have the same color and shapes or are they different?

MATERIALS

- A slice of moldy bread
- A piece of moldy cheese
- Stereoscope or a magnifying glass
- Paper
- Pencil

PROCEDURES

Observe the color and shape of the mold growths on both the bread and the cheese by looking at them under a magnifying glass or a stereoscope. The mold will look like a hairy, matted mass, called the *mycelium*. Notice the color, texture, and overall shape. Do any of the growths appear higher (i.e., taller from the surface of the food) than the others? Draw the shape of the individual mold filaments, called *hyphae*. Do you see any spore growth (i.e., small, round forms on stalks, called *sporangium*) that are rising above the surface of the growth? Draw any spore growth you see.

CONCLUSIONS

Is the color, texture, or shape of the filaments or the spores different on the different foods? What do you conclude about the physical appearance of the molds on bread compared with those found on the cheeses? Can you determine from your observations whether breads and cheeses are invaded by the same or different types of microbes?

GOING FURTHER

Collect different cheeses with different molds by going to a deli or grocery store and asking them to save you any moldy cheeses. Can different cheeses have different types of mold growth?

13

Milk, microbes, & more

Comparison of microbes in various types of milk products

Many microbes find milk and other diary products an excellent place to call home. Milk sours and then curdles when microbes are present and begin to reproduce. Pasteurization is a process that kills most of the microbes found in milk, making it safe for us to drink. In the days before pasteurization, raw milk may have carried tuberculosis, diphtheria, typhoid, or other dangerous disease-causing microbes. Today, pasteurized milk is safe, unless contaminated after the carton is opened.

Numerous types of bacteria affect milk differently. For example, some microbes sour milk and turn it lumpy. Others cause a sliminess or ropiness in milk, or gas bubbles that make the milk frothy. Still others can make milk translucent, or even turn milk red or blue. These color changes are rarely seen today because pasteurization prevents this from happening.

How does microbial growth differ between pasteurized and ultrapasteurized milk products? What type of reaction to microbial growth is seen in pasteurized versus ultrapasteurized dairy products, and how quickly does it occur?

MATERIALS

- Two clear, 1 cup glass containers
- Marker (that writes on glass)
- ½ cup of pasteurized half-and-half (fresh from a store; find out from grocery manager when new stock arrives)
- ½ cup of ultrapasteurized half-and-half (fresh from a store; find out from grocery manager when new stock arrives)

(If both pasteurized and ultrapasteurized heavy cream or light cream, are available, they can be used in place of the half-and-half.)

- Plastic wrap (e.g., Saran wrap)
- Two spoons
- pH paper

PROCEDURES

Sterilize the two glass containers following the procedures in chapter 1 "An Introduction to Microbiology." Label one sterilized glass "Pasteurized" and the other "Ultra." Pour ½ cup of each dairy product into the properly labeled glass. Loosely cover each glass with plastic wrap and keep at room temperature. Every 4 hours, observe the contents of each glass. To observe, gently swirl the contents with a spoon—are they clumped? Also, note the color and odor of the contents. You can lift the plastic cover for a moment to do this.

Use pH paper to measure the pH level in each container (see Fig. 13-1). Continue observing and reading the pH of both dairy products until both have noticeable growth.

13-1
Test the pH of the milk by holding the litmus paper in the glass.

CONCLUSIONS

What can you conclude about the effects of pasteurization and ultrapasteurization on microbial growth in dairy products? What type of reaction occurred in each glass and when? How did the pH change as the dairy products spoiled?

GOING FURTHER

Inoculate the plates to see how many different kinds of colonies grow. Research exactly how milk is pasteurized and ultrapasteurized at a processing plant. Continue the experiment but include a nonpasteurized product, such as heavy cream. Study the microbes under a microscope and try to identify them.

14

Microbe meals

Using microbes to create yogurt & control taste

Microbes are used to make many types of foods, such as beer, wine, and cheese. Vinegars are often microbial products. Microbes can be used as additives to increase the protein content of foods for humans and animals. The type of microbe used to make a food can affect the taste of that food. The taste is also affected by the source of the microbe and the number of cells in the original inoculation. How does the source of microbes used to make yogurt affect how quickly a culture forms and what it tastes like?

MATERIALS

- Two sterile pint mason jars with lids (follow procedures in chapter 1 "An introduction to microbiology" on how to sterilize containers)
- Marker (that writes on glass)
- A cup of plain yogurt
- One quart of milk
- Stove
- Three-quart pot
- One packet of yogurt starter culture (from organic food or grocery store)
- Tablespoon measurer
- Cookie sheet
- Oven
- Oven mitt
- Cooking thermometer
- Spoon

PROCEDURES

Label one mason jar "Starter" and the other "Plain." Warm the container of plain yogurt to 70 degrees F by holding it at room temperature. Heat the milk to 180 degrees F (almost boiling) and then pour 1½ cups of the milk into each of the jars.

Cool the milk in the jars to 100 to 110 degrees F (if its too warm or too cool, no yogurt will be made). Now, stir a packet of the starter culture into the jar labeled "Starter" and stir 3 tablespoons of the plain yogurt (now at room temperature) into the jar labeled "plain." The yogurt will act as the culture starter source for this jar. Close both jars and place them on the cookie sheet in an oven preheated to 100 degrees F. Be sure to keep the oven between 100 and 110 degrees F.

After 1 hour, check the cultures every one-half hour. When the contents have the consistency of custard, the yogurt is done. Do not shake them to find out if they are done, just gently tilt the jars (see Fig. 14-1). If the jars are disturbed too much, the culture might not form. It can take up to 9 hours to turn custard-like, but will probably take considerably less time. Note the time that each jar becomes custard-like.

14-1 Gently tip the jar to see the consistency of the milk.

Once a jar has become custard-like, remove it from the oven while wearing an oven mitt and put it in the refrigerator. After the yogurt cools, taste each jar. If no yogurt forms, then the temperature of the milk might have been too hot or too cool, or there might have been antibiotics in the milk. You can try the experiment again using organic milk, which should not have any antibiotics.

CONCLUSIONS

Does one culture take less time than another to create yogurt? Do the yogurts taste the same or did the source of the microbes (i.e., those from the starter package versus those from the existing yogurt) affect the taste?

GOING FURTHER

Repeat this experiment using different brands of yogurt. Do you find different tastes? Save a few tablespoons of the yogurt made from the starter culture and make a new yogurt with it. Does it still taste the same or does it change?

Part four

The history & tools of microbiology

Microbiology as a science has always depended greatly on the development of tools to see into the invisible world of microbes. The most obvious tool, of course, is the microscope.

Even though the first living cells were observed in the 1600s, microscopes and the study of microbiology didn't really advance until almost 200 years later. Many of these advances were connected to important discoveries in medicine. The *germ theory* of disease (the idea that microbes caused many diseases) was based on work by early microbiologists. In 1900, 5 of the top 10 causes of death were due to microbes. In the 1980s, only one of the top 10 causes of death were due to microbes. Learning about microbes has helped save countless human lives.

15

The good old days

What we see now versus what they saw then (1680s vs. 1990s)

Many people think the more magnifying power a microscope has, the better the scope must be. But advances in microscopy came not with more power, but with better *resolving power.* Resolving power is the smallest distance between two objects at which those objects may still be seen as separate things. Below that distance, the two objects appear as one. Your eye has a resolving power of about 0.1 mm. Therefore, when you look at your skin you don't see the tightly packed skin cells that make up your skin, you just see your skin as a whole, complete object.

The vast majority of microbes are much smaller than 0.1 mm. Therefore, their existence was largely unknown to humankind before proof of their existence occurred with the first microscope. Robert Hooke made the first microscope in 1664 (see Fig. 15-1); he looked at slices of cork and slices of beets and noted that they were composed of cells. He compared the structures he saw to the rooms monks slept in (i.e., cells). A few years later, Van Leeuwenhoek greatly improved Hooke's early microscope design and found that the world was filled with tiny *animalcules,* later known to be bacteria and protozoa. Van Leeuwenhoek's microscope increased the resolving power of the human eye 300 times.

What is the smallest size you can see using the microscope that is available to you, and how does this compare to what Hooke could see in the 17th century? How does this compare to what modern-day microscopes can view (e.g., electron microscopes, scanning/tunneling microscopes)?

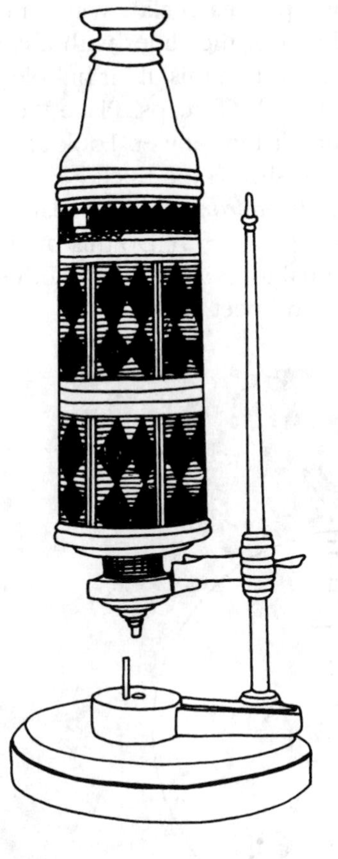

15-1 The first microscope was built by Robert Hooke.

MATERIALS

- Cork
- Beet
- Sharp knife
- Microscope slides
- Eyedropper
- Forceps
- Coverslips
- Microscope
- Ocular micrometer*
- Stage micrometer*

(* May substitute a thin, clear plastic ruler with millimeter divisions for these two items.)

PROCEDURES

Have your teacher or advisor use the knife to cut a thin slice of natural cork and a thin slice of the beet. The thinner the slice the better. Put each slice on a microscope slide. Place a drop of water on top of each slice with an eyedropper. Use the forceps or your fingers to slowly cover the slides with a coverslip. Try to prevent any air bubbles from forming as you do this. If air bubbles form, gently press on the center of the coverslip with the forceps. Place the slide under the microscope and focus on the cork under low power. Look at the edges of the slice. If light is not coming through, the slice is too thick.

Now measure the cells using an *ocular micrometer*, if available. This is an ocular with a ruler etched into the lens (see Fig. 15-2). Do this by first counting how many units in the ocular ruler are equal to the length or width of a cork or beet cell. Do this for a few different cork and beet cells.

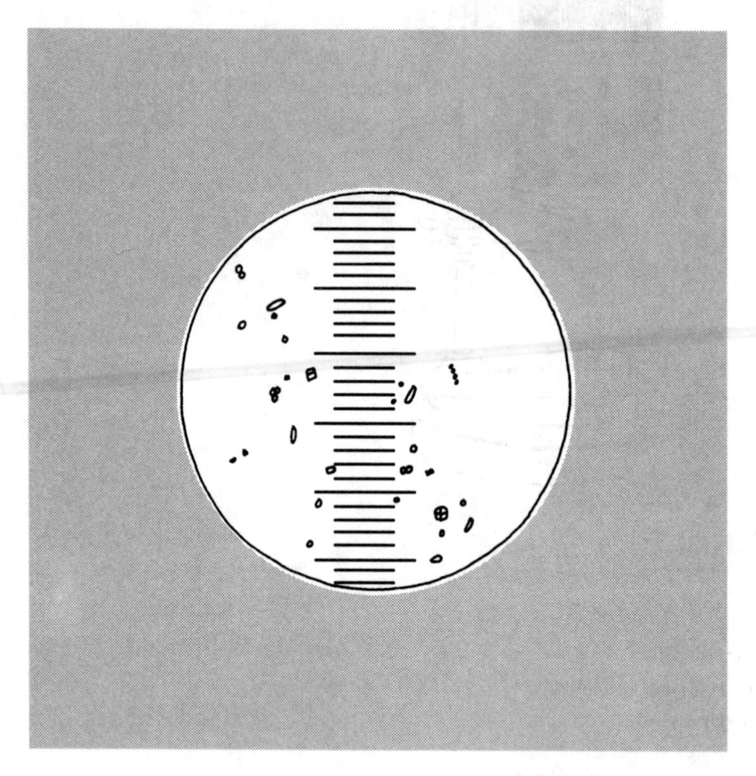

15-2 The ocular micrometer has a tiny ruler etched on the lens.

Finally, use the *stage micrometer*, which contains a tiny ruler, with very small divisions of a known size. Use the same microscope magnification that you used to measure the cells. The size of the divisions of the stage micrometer is usually etched on its glass (or its on a booklet that comes with the stage micrometer). Determine how many ocular micrometer units are equivalent to the stage micrometer units.

Once you have done this, you can determine the size of the cork and beet cells. Because you know how big the stage units are, and you know how many ocular units make-up a known number of stage units, you can calculate the size of an ocular unit. Furthermore, because you now know the number of ocular units that makes-up the width of a cork or beet cell, you know the exact size of the cell.

If an ocular micrometer and a stage micrometer is not available, use a thin, clear plastic ruler with millimeter measurement and lay it on top of the microscope slide and estimate the size of the cork and beet cells. This method is going to be a very rough estimate and will work only at the lowest magnification setting of the microscope (100×).

CONCLUSIONS

What is the average cell size for cork cells and beet cells? If you assume that this was the smallest size Robert Hooke could see, how does this compare to the smallest size that is possible to see today? Read about electron microscopy and scanning/tunneling microscopy to determine the smallest size that is visible with today's technology.

GOING FURTHER

Use cardboard or some other material to build scale models of a bacteria cell, an alga cell, a fungus cell, a protozoan, and a virus. Be sure they are proportioned properly. Then build a typical multicellular organism such as an ant for comparison.

16

Mr. potato head

The search for a solid growth medium

When you study large animals or plants, it is possible to observe a single individual. For example, you can measure how much one individual eats, what it eats, or how often she reproduces. You can also collect this information for many individuals and then determine the average for all the animals. What happens, though, when you can't observe a single individual? This is the major problem that microbiologists face.

The solution to this problem is to study a group of the same type of microbe. You must be sure, however, that the group you are studying is composed of all the same type of microbe—called a *pure culture.*

Groups of the same type of microbe can grow on a solid surface. These groups are called *colonies.* Colonies can have varying shapes, forms, and colors. The shape, form, and color of a colony help you identify the type of microbe that is present so you can learn more about it.

Colonies only grow on a solid surface. If you tried to grow microbes in a liquid, colonies won't form, although the microbes will reproduce. The first microbiologists cultured microbes on potato slices as the solid surface. They then switched to gelatin, which was later replaced by *agar.* Agar is currently the primary solid surface for microbial culture. Can you show why agar replaced gelatin as the solid growth medium in microbiology? What advantages does agar have over gelatin?

MATERIALS

- Four pint-sized mason jars with lids
- Four-quart pot (taller than the mason jars)
- Water
- Stove
- Marker (that writes on glass)
- Plain agar (available from scientific supply houses or some organic food stores)

- Plain, unflavored gelatin (available in grocery store)
- Scale or balance (to weigh 3 to 12 grams)
- Distilled water
- 100 ml measurer
- Two long-handled spoons
- Tongs
- 1-inch adhesive tape
- ½ teaspoon measurer
- Soil/hay infusion (To make a soil infusion, place in a mason jar a handful of recently cut grass and a handful of dead, dry grass or thatch from a lawn, garden, or abandoned field, and add about ¼ tsp. of soil. Add ½ cup water. Seal the jar and store it at room temperature for 2 weeks.)
- Thermometer

PROCEDURES

Sterilize the four pint-sized mason jars as described in chapter 1 "An Introduction to Microbiology." When sterilization is complete and the jars are cool, label them as follows: "Agar," "Agar+Infusion," "Gelatin," and "Gel+Infusion." Next make a 3% agar solution and a 12% gelatin solution in their respective jars. To do this, measure out 3 grams of agar and 12 grams of plain gelatin. Place each into the proper jar and add 100 ml of distilled water to each. Keep these jars standing up in the large pot containing hot water from the sterilization procedure performed earlier. Be sure the pot water line is slightly above the jar water line.

Put the pot on the stove and boil (see Fig. 16-1). Stir the contents of each jar with a spoon (use a different spoon for the gelatin and the agar jars). Let the solutions boil slowly for 15 minutes. Remove the jars from the pots with sterile tongs.

When the jars have cooled slightly, so there is no steam coming off of them, seal the jars without the infusion by screwing on their lids very tightly and then closing the lid's edges with adhesive tape to prevent evaporation. To complete the two jars with infusion, inoculate each with the ½ teaspoon of soil/hay infusion broth that is at least 2 weeks old. These jars should then be sealed as were the jars without infusion. Leave all the jars at room temperature.

CONCLUSIONS

When the jars without infusion have cooled, compare the physical characteristics of the agar to the gelatin. Are they both as solid? Move these two jars to a higher temperature (35 degrees C) and watch their physical characteristics over 2 weeks. Use an incubator, if available, or place them under a lamp and maintain a constant temperature using a thermometer. Do they stay solid or do they liquefy over time?

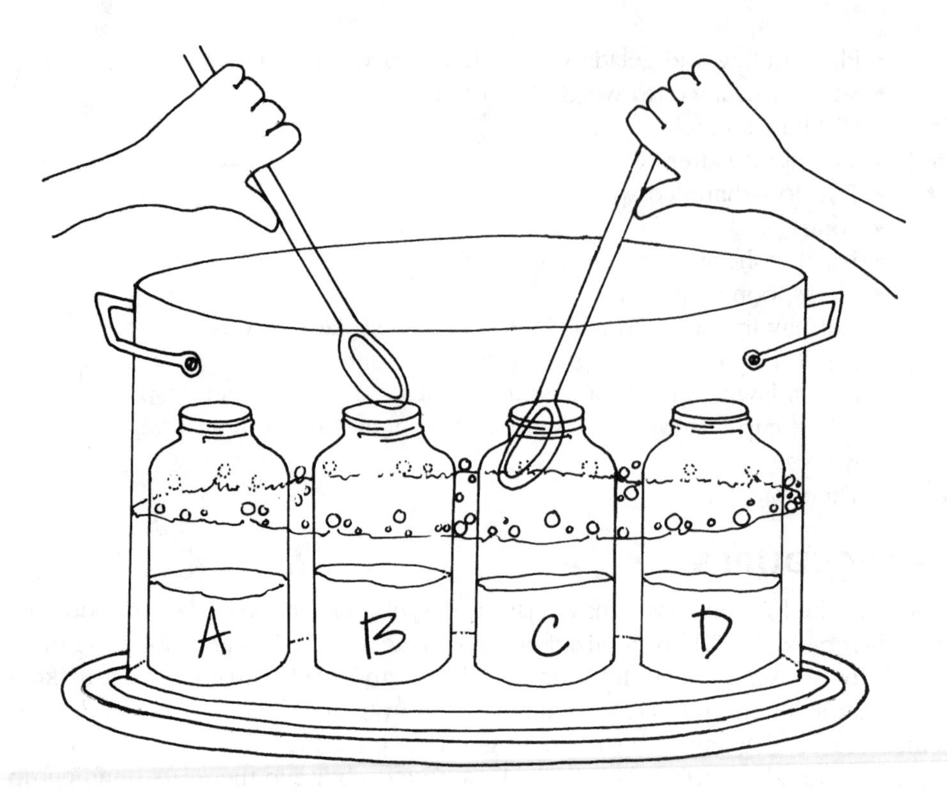

16-1 Be sure the water line within the jars is beneath the water line of the large pot.

Keep the jars with infusion at a temperature of 20 to 22 degrees C for 6 to 8 weeks. Observe the jars each week. Compare the results of the agar+infusion jars to the gelatin+infusion jars. Do they stay solid or do they liquefy over time? Why? What do you conclude about gelatin versus agar as a solid support medium?

GOING FURTHER

Early microbiologists used potato slices as a growth medium. Compare potato slices to gelatin and agar as growth media? Investigate the many special types of growth media available from scientific supply houses. What are they for and how do they differ?

17

I'm boiling, but am I clean?

Methods of sterilization

Microbiology, the study of microorganisms, depends not only on the microscope, but also on *sterility*—the absence of microorganisms. The development of microbiology has been tied to the development of sterilization methods. Sterile glassware, tools with which to inoculate, and especially media on which to grow the cultures are all extremely important instruments in microbiology.

Today, we have the *autoclave* (a pressurized steam bath) that sterilizes these tools, as well as other techniques, such as filter sterilization that filters out microbes. However, at the end of the nineteenth century, when microbiology was in its infancy, these devices and techniques were not available.

In 1864, Louis Pasteur understood that heat could kill microbes. Low heat levels slowed or eliminated the spoilage of wine to vinegar by killing the bacteria that produced the vinegar. This method is now known as pasteurization. Pasteur showed that boiling fruit juice once for 15 to 20 minutes would kill the microorganisms in the juice.

But this heat doesn't kill all bacteria. Some bacteria have a heat-resistant spore that is not killed after a single boiling. Another microbiologist at that time, John Tyndall, sterilized his liquid media by boiling them for 15 to 20 minutes on three consecutive days. This method became known as tyndallization. Can you find spore-forming bacteria in nature by comparing the sterilization methods of Pasteur and Tyndall?

MATERIALS

- Marker (that writes on plastic or glass)
- Vials containing sterile tryptic soy broth (12 to 18; available from scientific supply house)
- 3 teaspoons of garden soil
- Handful of hay or dried grass, shredded into 1-inch pieces
- Two (preferably three) pint-sized mason jars

- 4-quart pot
- Water
- Stove
- Test tube holder
- Sterile cotton
- 1-inch adhesive tape
- Incubator (or warm area in room under lamp)
- Thermometer
- Sink with hot water
- Soap

PROCEDURES

Sterilize the vials or test tubes as outlined in chapter 1 "An Introduction to Microbiology." Label three of the test tubes that will receive a single boiling (Pasteurization) as: "PlainP," "SoilP," and "HayP." Label three other test tubes to be sterilized over three consecutive days (Tyndallization) as: "PlainT," "SoilT," and "HayT." Add ½ teaspoon of fresh garden soil to both tubes marked "soil," and add a few pieces of hay (or dried grass) to both tubes marked "hay." It is preferable to make two or three sets of the six test tubes mentioned above.

Stand the mason jar up in the large pot and fill jars ⅓ to ½ full with water. Fill the pot with 1 to 2 inches of water. Place the six labeled test tubes in the mason jar and boil for 30 minutes. The mason jar holds the test tubes straight up in the boiling water (see Fig. 17-1). After boiling, remove each tube using the test tube holder. Stuff each tube with sterile cotton to prevent contamination from the air

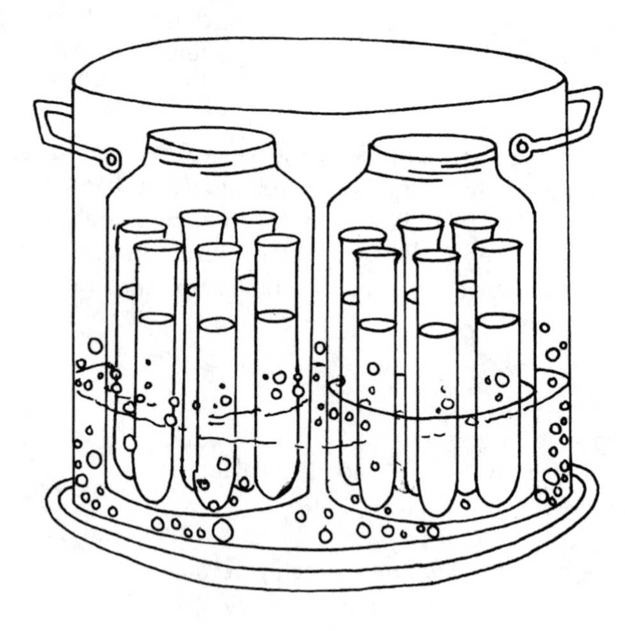

17-1
Place the six test tubes into the jars as they rest in the large pot.

(be sure your hands are well-scrubbed and clean). Seal the test tubes with adhesive tape. Do not allow the cotton to touch the broth.

Set the tubes marked with a "P" in an incubator, if available, at 37 degrees C or keep them in a warm spot in the room, preferably under a lamp. Note the temperature of the area with a thermometer and check it over time to be sure it doesn't fluctuate more than 2 degrees. Observe the tubes each day, looking for cloudiness, which indicates microbe growth. If the cotton gets wet, it might permit contamination from the air. Write down your observations.

For the tubes marked with a "T," boil them again for 20 minutes the next day and the day after that in the same fashion as performed earlier. However, be sure the tubes cool down to room temperature for at least 18 hours before each boil. If the cotton gets wet after the boil, replace with fresh, sterile cotton. Leave them at room temperature between each boiling.

When they have been boiled three times, place them in the incubator or the same warm location along with the tubes marked "P." Observe them each day to look for growth and take notes on their appearance as you have been doing with those marked "P."

CONCLUSIONS

Do you get growth in any or all of the tubes? If you only get growth in some tubes, what does this tell you about controlling spore-forming bacteria? Where are these bacteria found? If you get growth in all of the tubes, including the plain tubes, repeat the procedure taking care that the cotton is sterile and dry and the tubes are not open to the air except when inoculated.

Follow your teacher's directions about how to dispose of the cultures and equipment from this project. Gently dump out the contents of each tube down the drain while hot water is running in the sink. Be careful not to splash the contents around the sink or around the room. If you do, wash the area immediately with hot, soapy water. Rinse out the tubes with hot water, then dispose of them properly.

GOING FURTHER

Read about spore forming bacteria and how some can cause human diseases? Research the latest techniques used to sterilize foods.

18

Acid indigestion—microbe-style

The acidity of food vacuoles in a paramecium

Stains are used to detect the structure of microbes and determine what they are composed of. Some stains change color when the acidity of their environment changes. This type of stain can give information about the microbe's environment, as well as the microbe itself.

Congo red is a stain that is yellowish-red under basic conditions, cherry red under slightly acidic conditions, and blue under strongly acidic conditions. It can be used to stain the nucleus of a cell and to see the acidity of the food stored in the bodies of some microbes. Many *protozoans* (single-celled organisms), such as the paramecium, bring food into their body through an oral groove (opening). The food (in the form of *food vacuoles*) float throughout their body like liquid-filled balls. Different food vacuoles might have different acidities, depending on the types of food available and the eating habitats of the microbe. Furthermore, different food vacuoles could have varying acidities if digestion affects the pH level.

Can you determine the pH of different food vacuoles found throughout a paramecium. Do all the food vacuoles have the same pH or do they differ? Does their distance from the mouth of the paramecium affect the pH levels of the vacuoles?

MATERIALS

- Vaseline petroleum jelly
- Coverslips
- Congo red solution (available from scientific supply house)
- Live paramecia (You can purchase a paramecium culture from scientific supply house or collect your own from the scum on ponds or among

plants and debris at the edge of shallow ponds. To collect your own, go to any area with wet, decaying organic matter that has plenty of oxygen and scoop up some scum or plant material with a little water. Place it in a mayonnaise jar and seal the jar until you get home, then loosen the lid. Put the jar in a window that gets some *indirect* sunlight (direct sun will heat up the jar and kill everything) and leave it for a day or two. Then, take an eyedropper and pick up a few drops from the top layer of the jar.)

- Microscope depression slides
- Microscope
- Drawing paper
- Colored pencils (red, blue, yellow, black)

PROCEDURES

Smear Vaseline on the palm of one hand. Then, with your other hand, carefully put a rim of Vaseline on each edge of the coverslip by sliding each edge into the Vaseline on your palm (see Fig. 18-1). Next, with the Vaseline-side of coverslip pointing upward, place a small drop of neutral red solution in the middle of the coverslip. Let the drop air-dry.

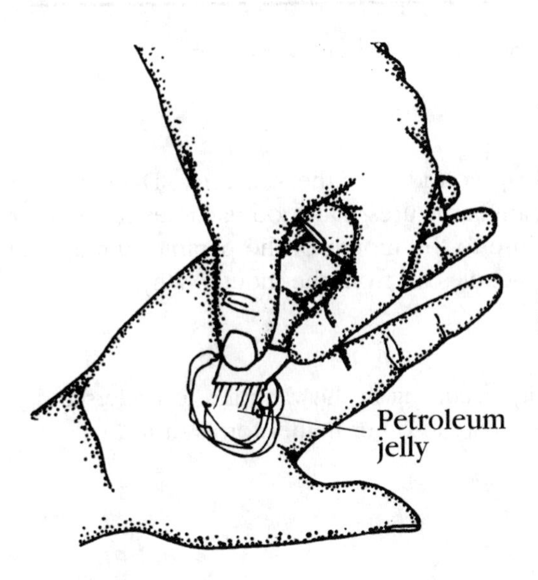

18-1 Rub the edges of the coverslip into the Vaseline that is on the palm of your hand.

Next, place a single drop of the paramecium culture on the center of the coverslip directly on the neutral red solution. With the coverslip still, Vaseline side up, place the depression slide over the coverslip and carefully press down (see Figs. 18-2 and 18-3). Then turn the slide over. This is called a *hanging wet mount.* Observe the protozoans under the microscope (100× magnification should be fine, unless you have very small, field-collected protozoans). Draw any structures that you see, using the appropriate color to represent the stains.

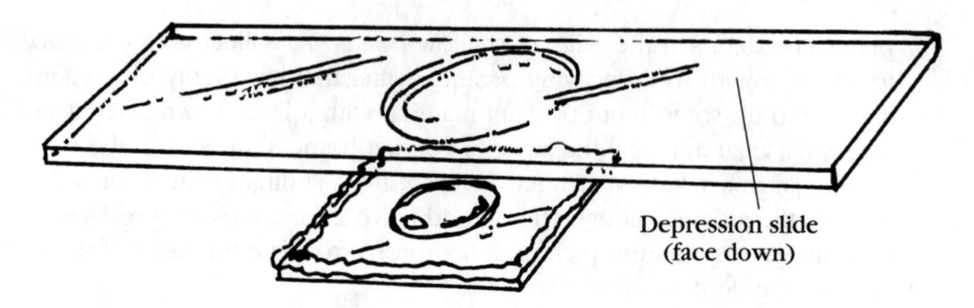

18-2 Press the depression slide face down onto the coverslip as shown.

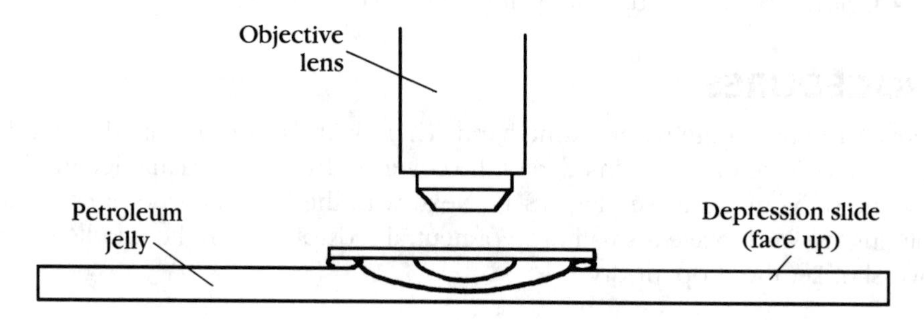

18-3 The resulting slide is called a hanging mount.

CONCLUSIONS

Use a book on protozoans to help you identify the structures. Does the Congo red solution change to blue in any structures? Do food vacuoles have different pH levels? Do food vacuoles close to the mouth of the paramecium appear to be a different color from food vacuoles far from the mouth? Why?

GOING FURTHER

Research the biochemical activity occurring in the vacuoles to understand your results. Compare this digestive activity to other higher forms of life.

19

Purity

Creating a pure culture

Microbiologists usually study pure populations of microbes in which all the cells are the same type. Observing the responses of the group (e.g., to a stimulus such as light) gives you the average response of all the individual cells in the population. Techniques that allow scientists to create these pure populations are important to microbiology.

Populations of the same type of microbe living on a solid surface, such as an agar plate, are called *colonies*. The shape, form, and color of a colony helps you to identify the type of microbe present and to determine if you have a pure colony.

Creating a *pure colony* (also called *culture*) requires using a *sterile technique* so you don't contaminate your culture with airborne microbes. *Sterility* is defined as the absence of life. Sterile technique is simply a procedure to eliminate all other microbes except the microbe you are interested in studying. Is it possible to make a pure culture of one kind of microbe on an agar plate by using a sterile inoculating loop?

MATERIALS

- Disinfectant solution (Consisting of 1 part bleach to 100 parts water, plus 1 part dish detergent)
- Sponge
- Rubber or latex gloves
- Paper towels
- Twelve sterile inoculating loops (or follow procedures in chapter 1 "An Introduction to Microbiology" on how to sterilize inoculating loops)
- Three or four sterile nutrient agar plates (available from scientific supply house)
- A mixed culture of microbes in a liquid medium (available from scientific supply house; or take ½ cup of soil and add some dead grass and 1 cup of water, let sit for one week)
- Incubator (or warm area in room under lamp)
- Thermometer
- 1-inch adhesive tape
- Notebook and pencil

PROCEDURES

Do these procedures in a room with very little air movement. This project has two sections. In the first, you will begin to isolate (separate) a single type of microbe on an agar plate from a mixed culture that you purchased or created. In the second section, you'll use the colonies that grew on the agar plate to create a second agar plate containing (hopefully) a pure culture.

To prepare for this project, squirt some of the disinfectant solution onto the surface of the table you'll be working on (do not use a wooden table) and wipe the disinfectant over the work surface with the sponge. Cover your hands with the latex gloves and squirt some of the disinfectant onto your gloved hands. Rub the disinfectant over your hands. Dry them with clean, paper towels.

Begin the first section by dipping a sterile inoculating loop into the mixed culture. Immediately close the mixed culture container. Move the loop around in the air as little as possible and don't set it down on the table. Open one of the agar plates very slightly, just enough to be able to get the loop in.

To begin moving the loop over the agar (called *streaking*), imagine that the plate is clock. Gently touch the loop to the 12 o'clock position and move the loop back and forth across the surface of the agar, at the top quarter of the plate (see Fig. 19-1, Step 1). Let the loop lightly touch the agar surface as you move it along. Then close the plate. Now, use another sterile loop. Once again, open the plate and move the loop in a straight line down (at a right angle) from the first few streaks (see Fig. 19-1, Step 2). Remove the loop and close the plate again.

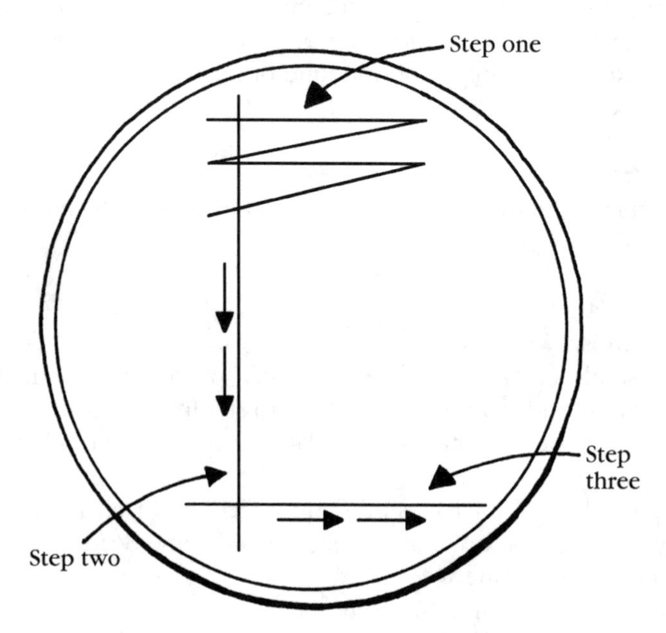

19-1 Follow each step closely to create a pure culture.

Finally, using another sterile loop, move the loop horizontally through the line you just streaked, so that the loop moves across this line and the plate as seen in Fig.19-1, Step 3.

If you dig into the agar plate with the loop, you must start over. It might take some practice before you are able to gently move the loop over the surface of the agar without tearing or digging into the surface.

Once all the streaking is completed, place the plate in an incubator, if available, at 37 degrees C or keep it in a warm area in the room under a lamp. Note the temperature of the area with a thermometer and check it over time to be sure it doesn't fluctuate more than 2 degrees. Let it incubate for 48 hours. Observe and make notes of the number, placement, and shape of the colonies. Repeat this entire procedure on two or three other additional plates.

If you had a good streaking technique, the fewest colonies will be in the section of the plate that you streaked last. Are there any colonies that stand alone, without touching any other colonies? You can now use one of these colonies as the source for the next culture.

To begin the second part of this project, open the plate and gently touch a sterile inoculating loop to the surface of the isolated colony (from Fig.19-1, Step 3, streak in the illustration). Remove the loop, close the plate, and streak the loop containing microbes from the first plate onto a fresh agar plate, exactly as you did in the original plate. Incubate this plate for 48 hours. Observe and make notes of the number, placement, and shape of colonies on this plate.

CONCLUSIONS

Do all the colonies on the second batch of plates appear the same? Do you have a pure culture? Do these colonies match the form and color of the original colony with which you started? If so, you have created a pure culture. Draw the shapes of the colonies that you find. If there are growths that appear "hairy," you might have fungal growth. Fungi release airborne spores with any air movement. It is best not to open plates with fungal colonies.

When you are done with a plate, seal its edges with masking tape and discuss with your teacher or advisor how to dispose of it. This is especially important for plates with fungal spores.

GOING FURTHER

How many different colony types can you isolate from your original mixed culture? Research the history of isolating microbes and the role this has played in our lives.

Part five

The world of bacteria

Bacteria are ancient. They are believed to be the first form of life on the planet Earth. Bacteria are found all over the world: underground, at the bottom of the sea, in hot springs, in Antarctica, almost everywhere. They are a very successful group of organisms, able to adapt to an extraordinary range of conditions. Bacteria usually reproduce by *fission*, which means they simply split from one cell into two *daughter cells*. The daughter cells are often smaller than the original cell. The speed at which they increase in cell size is dependent on the amount of available food and other nutrients, the temperature, the amount of moisture, and other factors in their environment. When the daughter cells have grown enough, they too will undergo fission. Some of the projects in this section investigate how bacteria grow.

Genetics is the study of characteristics that are passed from generation to generation. Passing characteristics from one generation to the next is called *heredity*. The smallest unit of heredity is called a *gene*. Genes often change, creating new versions of the organism (called a *mutation*). The new version might not function as well as the old, or it might function better. *Beneficial mutations* are those that help a cell adapt to new conditions. For example, many bacteria are now *resistant* to antibiotics that used to kill them. The development of resistance is a beneficial mutation for the bacteria. It lets the bacteria survive in a habitat where it was previously unable to survive. Some of the projects in this section explore bacterial genetics and mutations.

Bacteria are classified according to their shape, genetics, metabolism, among other methods. Correct classification is important to properly treat disease-causing microbes (*pathogens*), to be sure microbes used in our foods are the right ones for the job, and just to be able to accurately study the biology of the microbe. Some of the projects in this section investigate how bacteria are classified.

20

Snug as a microbe in a rug

Temperature & control of bacterial growth

All living creatures have an optimum temperature for their metabolism. *Metabolism* is a term that refers to all the biochemical reactions that occur in living organisms. Everything from digesting food to producing wastes are part of an organism's metabolism. The temperature of an organism's environment is very important because it controls the speed of all these reactions—called the *metabolic rate.*

Some complex multicellular animals, such as mammals and birds, have the ability to regulate their own body temperature, keeping their bodies at their optimum temperature. Bacteria and other organisms, however, are unable to control their internal temperature and are at the mercy of their environment. How does temperature affect the growth rate of a bacteria?

MATERIALS

- Two test tubes of sterile tryptone broth (available from scientific supply house)
- Two coolers
- Cold water (less than 15 degrees C)
- Ice
- Two thermometers (5 to 50 degrees C)
- Warm water (greater than 35 degrees C)
- Two wire test tube holder racks (that can fit into the bottom of the coolers)
- Two wooden spoons
- Marker (that writes on glass)
- 14 microscope slides

- Microscope with immersion oil lens
- Clock
- Culture of *streptococcus faecalis* (available from scientific supply house)
- Sterile pipettes (graduated in 0.1 ml markings; follow procedures in chapter 1 "An Introduction to Microbiology" on how to sterilize pipettes)
- Coverslips (22 mm × 22 mm)
- Stage micrometer

PROCEDURES

Label one of the test tubes containing the sterile tryptone broth "Cool," and the other "Warm." Fill one of the coolers with 2 to 3 inches of cool water and adjust the temperature to 15 degrees C with cold water or ice (use thermometer). Fill the other cooler with hot water and adjust the temperature to 35 degrees C.

Put the labeled tubes into their appropriate water bath by standing them up in a test tube rack at the bottom of the cooler. You'll have to stabilize the rack on the floor of the cooler (see Fig. 20-1). Allow enough time for the tubes to reach the same temperature as the water bath. Keep a thermometer and a long, wooden spoon in each bath and add water as needed to keep the temperature stable.

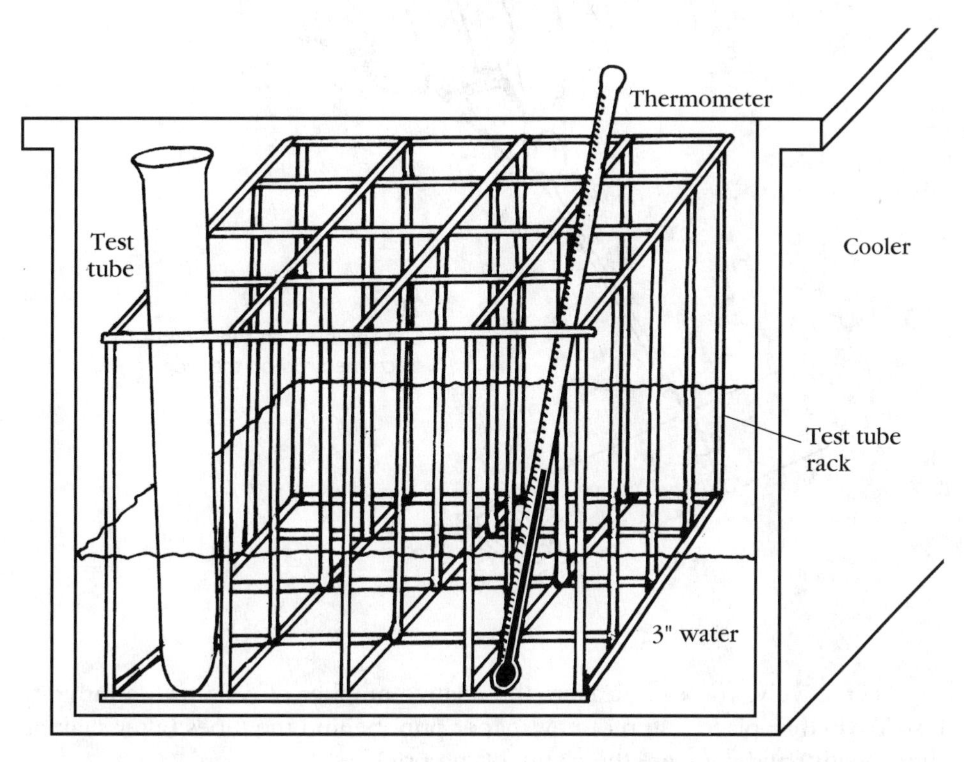

20-1 Rest the test tube rack at the bottom of the water bath as seen here.

When you add water, stir it around to be sure the bath has the same temperature throughout. This is especially important if you are using ice to cool the water.

Label the microscope slides to show the temperature ("C" for cold and "W" for warm) and how long the tube was in the bath (0 to 4 hours). Label the slides as follows: "C-0," "C-.5," "C-1," "C-1.5," "C-2," "C-3," "C-4," and "W-0," "W-.5," "W-1," "W-1.5," "W-2," "W-3," "W-4".

Now add 0.1 ml of fresh growing culture of streptococcus faecalis to each tube in each of the coolers and gently shake to mix. Put the tubes back into their water bath racks. You might need some assistance performing the remaining procedures. Then use a sterile pipette to take a 0.1 ml sample out of each tube. Return the tubes to the racks in the water bath.

Place the first sample in the middle of the microscope slide labeled "C–0" (see Fig. 20-2). Cover the drop of culture with a coverslip and, using a microscope, count the number of cells in the sample using a stage micrometer. (See chapter 1 "An Introduction to Microbiology" for instructions about counting cells.)

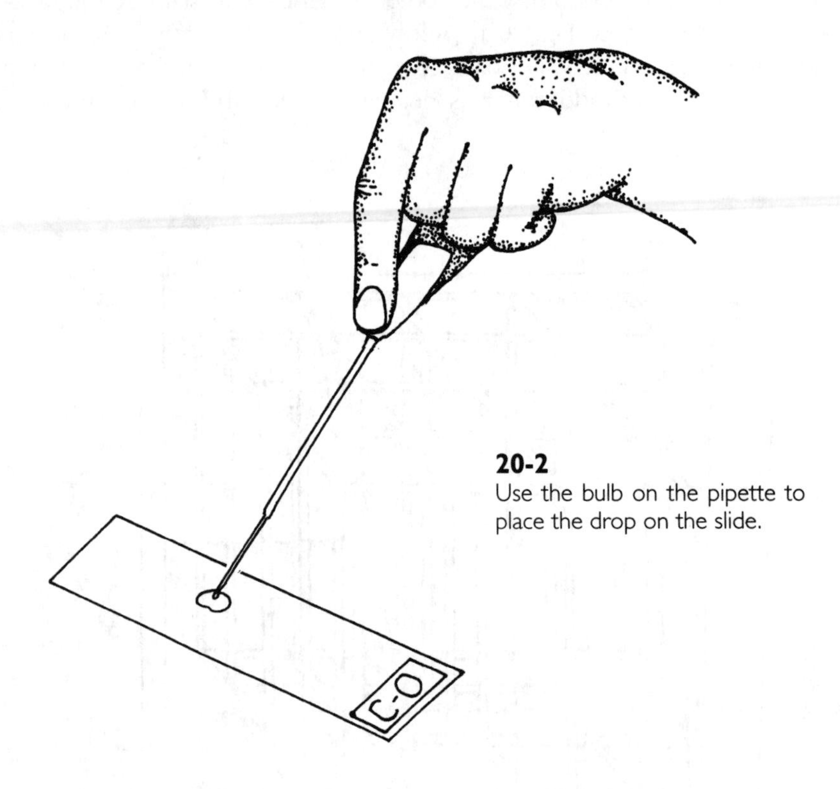

20-2
Use the bulb on the pipette to place the drop on the slide.

Alternatively, you can measure the relative number of cells with a turbidity test. To do this, place a strip of newspaper print behind the tubes (after shaking them gently) and compare the clarity of the print.

Using one of these techniques, measure the "W-0" tube. This is the number of cells at time "0." After 30 minutes has passed, gently shake each tube and take another 0.1 ml sample.

Immediately return the tubes to the water bath. Place the sample on the slide marked "C-.5," and repeat the procedure as explained above. Do the same for the tube marked "W-.5." This is the number of cells for the 30-minute mark. Repeat this procedure for each time interval up to 4 hours.

CONCLUSIONS

Graph the number of cells per 0.1 ml culture by time for the two sets of tubes. How do they differ? Do they differ throughout the entire time period or only in portions of it? What do you conclude about the effect of temperature on the growth of these bacteria?

Discuss with your teacher or advisor the best method of disposing the cultures and disinfecting the glassware. Be sure to disinfect all surfaces after using this culture and carefully dispose of all contaminated materials, including the microscope slides.

GOING FURTHER

What would happen if the temperature was hotter or colder? What is the temperature tolerance range for these or other bacteria?

21

Hippie bacteria

The effect of a black light on bacteria

The sun is the source of energy for all the living creatures on our planet. While the sun produces *visible light*, it also produces energy invisible to our eyes, such as *ultraviolet (UV) light* and *infrared (IR) light*. UV light can damage some biochemicals. *DNA*, the molecule that holds the genetic information of life, can be damaged by UV light (*radiation*). This damage can lead to many serious problems, including mutations.

The ozone layer is very important in absorbing much of the sun's UV radiation. Loss of the ozone layer leads to greater amounts of UV light reaching the Earth's surface, resulting in plants and animals (including humans) with more damaged DNA. Does UV light affect a bacteria's ability to grow?

MATERIALS

- A germicidal lamp, or a 20 to 40 watt UV (black light) lamp (a germicidal lamp is safer and might be available at a high school)
- Cardboard box (tall and strong enough to hold the UV lamp)
- Rubber gloves
- Sterile nutrient agar plates (available from scientific supply house)
- Marker (that writes on glass)
- Pure, liquid bacteria culture (e.g., *streptococcus faecalis, rhizobium leguminosarum*, or other species recommended by your teacher)
- Sterile pipettes (follow procedures in chapter 1 "An Introduction to Microbiology" on how to sterilize pipettes)
- Sterile spreading rods (follow procedures in chapter 1 "An Introduction to Microbiology" on how to sterilize spreading rods)
- Watch with second hand
- 1-inch adhesive tape
- Incubator (or warm area in room under lamp)
- Thermometer
- Notebook and pencil

PROCEDURES

Set up the UV light (or germicidal lamp) so that it will shine down from the top of the cardboard box. *Warning*: Do not have the bulb close to the edges of the box or the cardboard might catch fire. Do not look directly at the light because UV light can damage your eyes. Also, don't expose yourself to the UV light for extended periods of time. If you put your hands under the box while the light is on, be sure to use gloves to protect your skin.

Label each agar plate with one of the following: "0," "30 seconds," "1 minute," and "2 minutes." Add 0.1 ml of the pure liquid culture to the center of each plate using the sterile pipette. Use a sterile spreading rod to push/spread the culture equally over the surface of the plate. Cover the plates. Do not set the spreading rod down on the work area because it will become contaminated.

After the plate dries (there will be no visible moisture on the surface of the plate), expose the "30 seconds" plate to UV light for 30 seconds by taking off the lid of the plate and holding it in the box beneath the UV light for the correct amount of time (see Fig. 21-1). UV light does not penetrate through solid objects very well, so it is necessary to take the plate covers off. Hold the plate 5 centimeters away from the light. Wear gloves for this step. Repeat this procedure for the other plates for the proper duration.

21-1 Hold the petri dish in the box beneath the UV light.

Once a plate has been exposed to the UV, tape the plate closed by sealing the edges with adhesive tape. Place all the plates in an incubator, if available, at 37 degrees C or keep in a warm area in the room under a lamp. Note the temperature of the area with a thermometer and check it over time to be sure it does not fluctuate more than 2 degrees. Count the number of colonies present on each plate every day for 1 week. Make daily notes on the number and size of colonies on each plate.

CONCLUSIONS

Create a graph that plots the growth curve for each plate. How does the curve of the control plate ("0") compare with the others? What do you conclude about the effect of UV light on the growth of bacterial cells?

GOING FURTHER

Repeat this experiment, but keep the lids on the plates. Do you get the same results? Repeat this experiment with a spore-forming bacteria. Does UV light affect the growth of these cells?

22
Miniature movement
Temperature & Brownian movement

Many microbes, including some bacteria, can move around on their own. This helps them reach food or get away from places that might be dangerous. When you see a microbe moving about under the microscope you might think you are seeing the movements of a living creature, but this is not necessarily true. What you are probably seeing is something called *Brownian movement.* This is movement caused by molecules in the water striking the microbe and shaking it.

Molecules of all kinds don't just sit still. They are constantly moving. This causes liquids and gases to spread out (*diffuse*) throughout an area. Diffusion is the movement of molecules from a region of high concentration to a region of low concentration. Does temperature affect Brownian movement?

MATERIALS

- Water
- Two eyedroppers
- Two microscope slides
- India ink
- Coverslips
- Microscope
- Notebook and pencil
- Freezer or ice
- Lamp

PROCEDURES

Put a drop of water on a microscope slide. Add a tiny drop of India ink to the water. Cover the drop with a coverslip, taking care not to trap any air bubbles under the coverslip. View the drop with a microscope under 100× and then 400× magnification. Take detailed notes of any motion that you see. Be sure the objective is not touching the coverslip. Now create a second slide similar to the first.

Put one slide in the freezer or on ice. Put the second slide under a lamp (see Fig. 22-1). After at least 5 minutes, view the slides under the microscope once again looking for movement. Take notes about the movement in each.

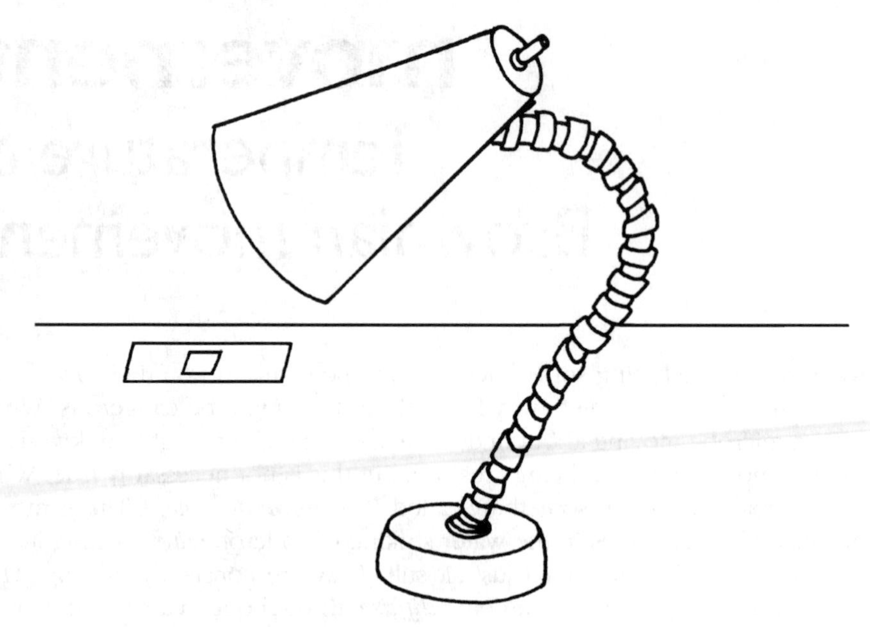

22-1 Warm the slide beneath a lamp.

CONCLUSIONS

How does the movement compare between the three slides (i.e., normal, cooler, and warmer temperatures)? Has there been a change in the amount or type of motion? What can you conclude about the effect of temperature on Brownian movement? How dramatic is this effect?

GOING FURTHER

Can you devise an experiment that will enable you to distinguish between true cellular motion and Brownian movement?

23

Milk, microbes, & classification

Using litmus milk to classify microbes

Because bacteria are so small and so diverse, it is difficult to classify them based only on their appearance. One of the ways to classify bacteria is based on what they do. For example, do they change the pH of the growth media? Do they use sugar? Do they produce gas? Can they survive in the absence of oxygen? *Litmus milk* is a growth media used to classify bacteria. *Litmus* is an indicator of pH (whether something is acidic or basic). Litmus is pink in the presence of acid, and blue to purple when neutral or basic.

Milk contains two possible sources of energy for bacteria to use. One is *casein*, which is the major protein in milk. This is what makes the milk opaque (cannot see through it). The other source of energy in milk is a sugar called *lactose*. Some bacteria use lactose for energy, while others use the casein.

If the bacteria use the casein for energy, the casein will break down making the milk become clear and changing the pH, which makes the litmus turn purple. If lactose is used by the bacteria for energy, *lactic acid* is formed, which turns the litmus pink. If enough acid is produced, the casein clumps up and curdles the milk.

How do different types of bacteria use milk? Which energy source does a *Lactobacillus* species and *Pseudomonas fluorescens* use?

MATERIALS

- Marker (that writes on glass)
- Six sterile test tubes of litmus milk media or create your own using litmus milk powder (both are available from scientific supply house)
- Sterile inoculating loops (follow procedures in chapter 1 "An Introduction to Microbiology" on how to sterilize inoculating loops)

- Culture of *Lactobacillus* species (available from scientific supply house)
- Culture of *Pseudomonas fluorescens* (available from scientific supply house)
- Incubator (or warm area in room under lamp)

PROCEDURES

Label three of the tubes "Lactobacillus" and the other three "Pseudomonas". Using a sterile inoculating loop, inoculate each of the three tubes labeled "Lactobacillus" with *Lactobacillus* culture (see Fig. 23-1). Use another sterile loop to inoculate each of the *Pseudomonas* tubes with the Pseudomonas culture.

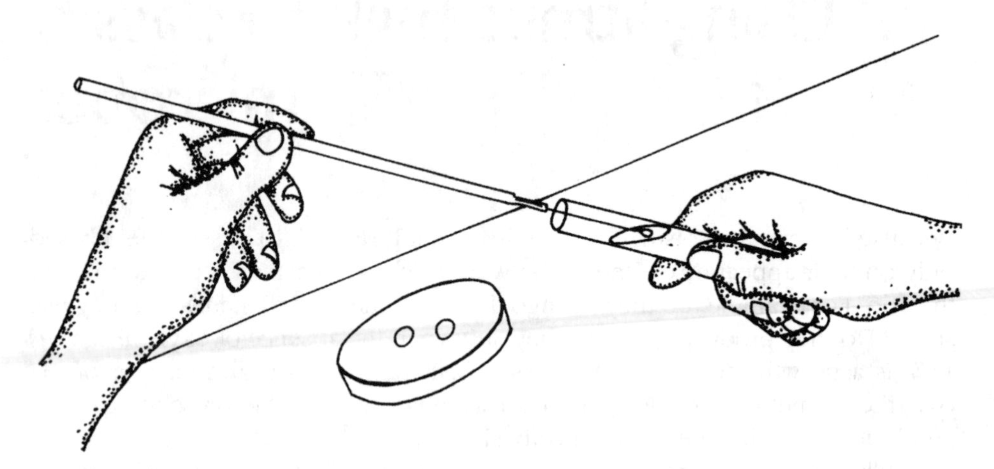

23-1 Use the inoculating loop to transfer the culture from the tubes to the agar plates.

Incubate all six tubes at 35 degrees C. If you don't have an incubator, place them in a warm area in the room under a lamp. Use a thermometer to monitor the temperature. Be sure the temperature doesn't fluctuate more than 2 degrees. After 24 hours, observe whether the tubes are opaque or clear, their color, the presence of layers, the presence of coagulation (clumping), and the presence of white solids at bottom of tube. Observe these characteristics once again after 48 hours.

CONCLUSIONS

What happened to the litmus milk in each set of tubes? The three tubes within each set should have similar results. If not, your sterile technique didn't work and the experiment must be repeated. How did the two sets of tubes differ? What can you conclude about which portion of the milk (protein or sugar) was being used by the *Lactobacillus* species and the *Pseudomonas fluorescens*?

Discuss with your teacher or advisor the proper way to dispose of the cultures and materials used in this project. When you are done with the tubes, carefully dispose of their contents down a drain with running hot water. Be sure to disinfect all surfaces after doing this experiment and dispose of all contaminated materials.

GOING FURTHER

Perform a similar experiment using other types of bacteria. What happens to the litmus milk?

Part six

Other microorganisms

The previous section contained projects about bacteria, but bacteria is only one of five major groups of organisms studied by microbiologists. The other four include algae, fungi, protists, and viruses. Projects in this section explore these other groups.

Algae are producers, which means that, just like green plants, they make their food by converting solar energy into chemical energy. Algae are very important to the production of oxygen on Earth. *Fungi* are important for a different reason. They are decomposers, responsible for breaking down and recycling dead organisms (*organic matter*) so they can be used by other organisms. Fungi can be roughly grouped into three categories: the molds, the yeasts, and the mushrooms.

Protists are a diverse group, which include organisms that produce their own food and others that must "eat" their food. The protists are the most complex of the single-celled organisms. They include single-celled organisms, such as amoebas, and ciliates, such as the paramecium, as well as the parasites that cause malaria. They live in many different types of habitats and have a variety of different lifestyles.

Another group that is even more difficult to classify are viruses. *Viruses* do not even fit the traditional definition of a living creature because they cannot reproduce on their own. They must have a host before they can reproduce. Viruses infect a host, and insert their genetic information into the host, which forces the host to reproduce the virus. Each and every species of organism on this planet probably has a virus that infects it.

24

One cell does it all

Classification of fungi, bacteria, algae, & protists

Microbe is a general term for all life that cannot be seen with the unaided human eye. But microbes can be divided into five categories: bacteria, algae, protists, fungi, and viruses. This division is based on their cell structures and what they need to survive. Can you learn about these different groups and explain how they differ? Collect examples of each type of organism (excluding viruses) from nature. Can you find them all?

MATERIALS

- Pond water or water from an old pond
- Damp soil from a garden, forest, or lawn
- Microscope slides
- Coverslips
- Eyedropper
- Water
- Microscope
- Drawing paper and pencil
- Book to identify microbes (a general microbiology text, such as those listed at the back of this book)

PROCEDURE

Collect small amounts of water from the surface of a small pond or old puddle and damp soil from a garden, forest, or lawn. Make a microscope slide from the water sample by putting a drop of the water on a slide and gently lay a coverslip over the top. Do this slowly so it doesn't trap air bubbles. For the soil sample, place a very small amount of damp soil on a microscope slide and put a drop of water over it. Use a toothpick to spread the mixture smoothly over the slide. Then place a coverslip over it.

View the slides under 100× magnification and then 400× magnification. When you find a microbe, draw its structure and shape. Try to get a rough estimate of the size of the microbe (see "Counting Cells" in chapter 1 "An Introduction to Microbiology" to see how this can be done). Use oil immersion if available.

Use the drawings and the identification manual to identify the types of organisms you find. Continue looking until you find an example of each type of microbe (i.e., bacteria, algae, protists, fungi). If you can't find a microscopic fungi, can you find a *macroscopic* (visible to the naked eye) type?

CONCLUSIONS

Were you able to find all four types of microbes? How did their shape and structure differ? How did the cell size compare with cells of another type? Did you find different types of cells in the water versus the soil? Draw the microbes to scale in a circle that represents the field of view.

GOING FURTHER

How many different types of each of the basic groups (i.e., bacteria, algae, protist, fungi) can you find in each habitat? Use your identification guides to help you categorize the different types. Are some types of microbes more common in certain habitats than in others?

25

They're everywhere!

Mold spores in our home

Molds are a type of fungi that consists of many cells (i.e., they are multicellular). The main body of a mold is made up of filaments that branch out into the substrate on which they are living (e.g., the substrate might be a rotting log that provides their food). The body of the mold is called the *mycelium*. Molds reproduce with *spores*, usually on filaments rising above the mycelium. The spores float through the air on the wind. If they land where there is enough moisture, they might grow to form a new mycelium.

These spores are released by the millions. Many humans are allergic to mold spores. Some cause respiratory irritation, while others can cause disease. Does your house contain mold spores? Yes! Every house does. But are these spores evenly distributed throughout your house?

MATERIALS

- Marker (that writes on glass)
- Sabouraud media agar plates (available from scientific supply house)
- 1-inch adhesive tape
- Incubator (or warm area in room under lamp)
- Thermometer
- Drawing paper and pencil
- Stereoscope or a magnifying glass

PROCEDURES

Label the sides of two Sabouraud plates "Kitchen." Open one plate in your kitchen and wave it back and forth through the air twice. Close the plate and seal it with adhesive tape. Do the same procedure for the other "Kitchen" plate.

Repeat this procedure in different areas of the house (e.g., in a refrigerator, in a bedroom, under a bed, in the basement, in the attic, in the living room). Label each set of two plates with the proper location (see Fig. 25-1).

25-1 Label two petri dishes for each room.

Place all the plates in an incubator, if available, at 37 degrees C or keep in a warm area in room under lamp. Note the temperature of the area and check it over time with a thermometer to be sure it does not fluctuate more than 2 degrees. Twice each day, note the shape of the colonies, their color, position on the plate, size of colonies, and the number. Draw pictures of these colonies to show their growth patterns. Use a magnifying glass or a stereoscope to see the colonies in more detail.

CONCLUSIONS

Do you find different types and numbers of colonies in the plates opened in different rooms? Do you find consistent types of colonies in the two plates from the same room? What can you conclude about mold distribution in your house?

GOING FURTHER

Perform a similar experiment but make the cultures during different seasons. For example, do it during the winter when the house has been sealed up to keep the heat in, and again in the summer when the air conditioner has been circulating air throughout the home. Is the distribution of mold affected by house air currents? Another possibility is to perform the experiment again in an energy-efficient building with very little outside airflow.

26

Hairy bread

Bread mold & preservatives

Fungi is a very diverse group of organisms. Molds and yeasts are different types of fungi. Some fungi are large and consist of many cells (i.e., multicellular), such as mushrooms. Others are single-celled organisms, as small as bacteria. Some fungi appear to look like plants, but they never contain chlorophyll and must consume their food as opposed to plants that can produce their food.

Fungi feed on a wide variety of substances including cellulose, which is found in most plants. Because of their ability to eat cellulose, some fungi feed on our foods, spoiling them. Bread is often spoiled by a group of fungi commonly called bread molds. Do preservatives successfully control bread mold growth?

MATERIALS

- Marker that writes on plastic
- Six plastic bags (Ziploc-type)
- Fresh bread with preservatives (purchased at a grocery store)
- Fresh bread without preservatives (purchased from most supermarkets or organic or health foods stores)

(Both breads must be the same type (e.g., wheat, rye) and must be fresh from the stores on the same day)

- Notebook, drawing paper, and pencil
- Magnifying glass or stereoscope

PROCEDURES

Label three plastic bags "Preserved." Expose the slices of bread to the air for 5 minutes. Then put one slice of the bread with preservatives into each of the three bags. Repeat this procedure with the bread without preservatives in properly labeled bags. Seal the bags.

Every one or two days, observe all the slices. Is there any visible mold growth? Note the date that growth first appears for each slice. Draw the appearance of the mold. Observe the growth under a magnifying glass or stereoscope.

Does the growth appear "mold-like" (see Fig. 26-1). Is it growing up and out of the bread as well as into the slice? Draw its appearance. Continue observations for a number of days after growth first appears.

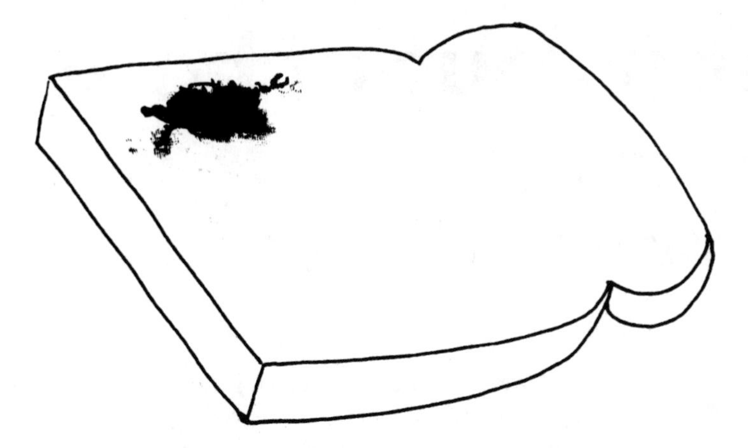

26-1 Look for bread mold growth on each slice.

CONCLUSIONS

Create a graph that represents the numbers and types of colonies found on the two breads. Over the course of the experiment, how effective are preservatives at controlling bread mold growth?

GOING FURTHER

Inoculate (transfer) some of the mold from bread without preservatives onto bread with preservatives. Will the mold grow on the new slice? How long until growth is seen? What happens when you use different types of bread (e.g., white, rye) or different brands of bread with no preservatives? How does temperature affect mold growth?

27

Let us out of here

Mold spores in a vacuum cleaner

Molds are all around us. Molds and other types of fungi release spores. Spores can be carried in air, water, or soil. When spores land in the right kind of environment, they begin to grow (*germinate*). If the environment has enough nutrients, new growth will expand and begin to develop new spores, to begin the cycle again. Are there mold spores in vacuum cleaner bags? What kind of environment is needed for the spores to germinate?

MATERIALS

- Twelve resealable plastic bags (Ziploc-type)
- Marker that writes on plastic
- Newspaper
- Vacuum cleaner bag (a used, filled bag)
- Scissors
- Teaspoon measurer
- Clean cloth rag, cut into 12 pieces (2 inches square)
- Water
- Dry food (dry cereal)
- Moist food (cereal with milk)
- Magnifying glass or stereoscope

PROCEDURES

Label three of the resealable plastic bags "Dry," three of them "Damp," three of them "Dry Food," and the last three "moist food." Lay a few sheets of newspaper on the floor. Lay the used vacuum cleaner bag on the newspapers. Cut the bag with the scissors (see Fig. 27-1) so you can insert the spoon into the bag. Place 1 teaspoon of vacuum cleaner bag contents into each of the plastic bags.

Soak three of the pieces of cloth in water. Ring them out, so they aren't dripping wet and put one into each of the bags labeled "Damp." Put one of the three

27-1 Lay the vacuum bag on the newspapers and make a small cut to remove a spoonful of its contents.

pieces of dry cloth into each bag labeled "Dry." Next, rub 1 teaspoon of dry cereal into another three (dry) pieces of cloth (you'll get some dry crumbs of cereal on the cloth) and place each of these into one of the bags labeled "dry food." Finally, rub the last three (dry) pieces of cloth into the moist food. Put these in the "moist food" bag. Seal all the bags and shake them well.

Incubate the bags in a warm area (30 degrees C) of the house for 48 hours. Then look at each bag and make note of any growth. Observe the bag contents with a magnifying glass or a stereoscope. Continue the experiment for another 2 days and repeat your observations. Continue looking at the bags every 2 days for a 2-week period.

CONCLUSIONS

Did you find mold growth in any of the bags over the 2-week period? If so, which bags had growth and when did it develop? What can you conclude about the presence of mold spores in the vacuum cleaner? What can you conclude about what condition is needed for the spores to germinate? Is food necessary?

GOING FURTHER

Does your vacuum cleaner catch all the spores it sucks in? Hold a damp rag over the vacuum exhaust, while it is running, for 2 minutes. Try to culture what collects on the rag.

28

Pull down the shade, turn out the lights

Protists, motion, & light

Many microbes can move. Some bacteria and protozoans have *flagellum,* which are like long tails. They use them to whip the water to provide propulsion. Many protozoans have *cilia,* which are short hairs that work like oars to move the cell around. Paramecium is an example of a *ciliated protist.* Do some protozoans move to or away from light?

MATERIALS

- Freshwater pond collection (take surface scum and some plant debris from a pond and place in mayonnaise jar with a lid)
- Aluminum foil
- X-acto knife
- Lamp with 75 watt bulb
- Eyedropper
- Microscope slides
- Coverslips
- Plastic wrap
- 1-inch adhesive tape
- Microscope
- Drawing paper and pencil

PROCEDURES

Wrap the jar containing the freshwater culture with foil. Keep the jar covered with the lid. With supervision from a teacher or advisor, take the X-acto knife and slice a 2–3 mm square window into the foil. Place a 75 watt lamp in such a way that light is shining into the square window (see Fig. 28-1). Remember not to put the light too close to the jar because if the contents of the jar heats up, your collection will die. Leave the light shining on the jar for 30 minutes. Then remove the foil without disturbing the jar. Note what you see in the jar? Use the eye-

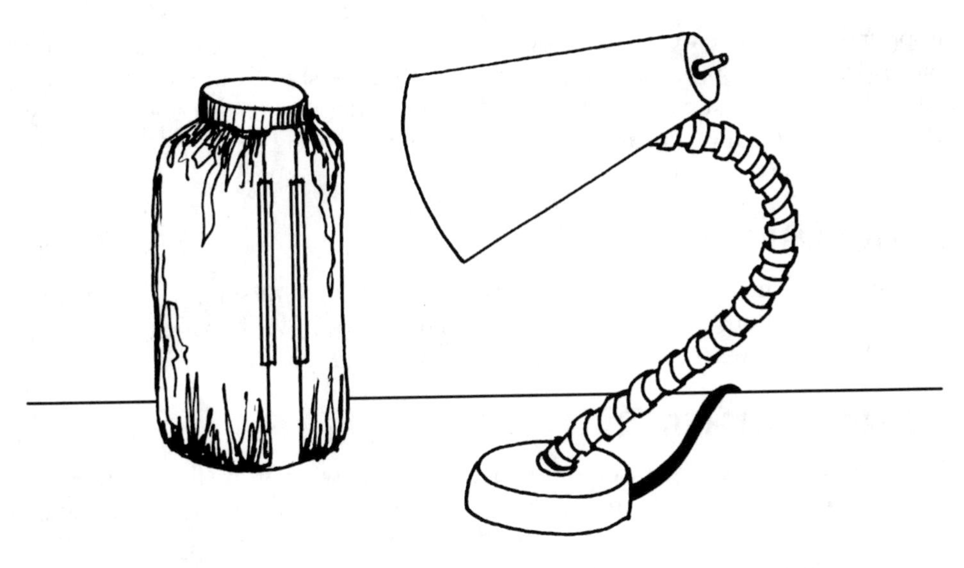

28-1 Leave a "window" of the jar open so light can pass through it.

dropper to take a collection from the window portion of the jar and put a drop of this sample on a microscope slide with a coverslip, labeled "light window."

Next, shake the jar and replace the lid with a clear, plastic film so light can shine down through the jar. Take a 5 mm square piece of foil and tape it to one side of the jar. Place the light directly in front of the foil strip (but once again, not too close so the water collection does not heat up). Leave the light in place for 30 minutes (see Fig. 28-2). Return and take another sample using the eye-

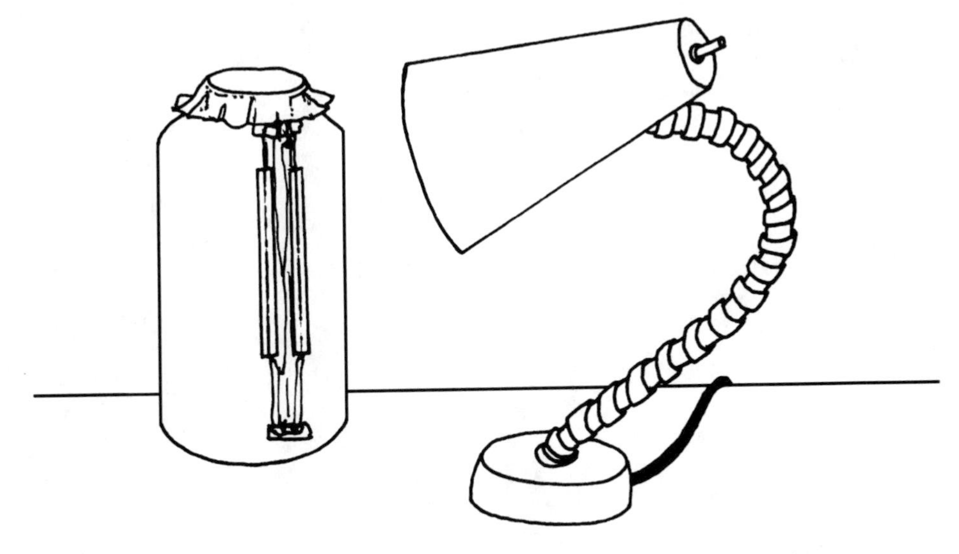

28-2 Place a piece of foil on the jar to block the light at this location.

dropper, but this time from the shaded portion of the jar (the area shaded by the foil strip). Put a drop of this sample on a microscope slide with a coverslip, labeled "shadow." Next, view both these samples under the microscope and draw what you see. Finally, repeat this procedure, but leave the jar in the light for 24 hours, before taking samples.

CONCLUSIONS

What types of microbes did you find in the three different samples? What can you conclude about the ability of some microbes to move? What do you conclude about their reaction toward or away from light?

GOING FURTHER

Research why these organisms moved toward or away from the light. How many different types of light-reactive organisms can you find in these or in other samples?

29

Keeping ahead of the competition

Predator-prey population relationships

Paramecia are complex single-celled organisms. They usually reproduce in the same way bacteria reproduce, by *fission*. Fission is the splitting or budding of one cell to form two cells. Paramecia are predators of bacteria, meaning they capture and eat bacteria for food. They ingest bacteria cells whole and slowly digest them.

If you were growing a culture of paramecia, you must also have bacteria in your culture for their food. If the population of predators (the paramecia) eats all the prey (the bacteria), then the paramecia will die also because they'll have no more food to eat. If the predators are to survive, the prey cannot be totally destroyed. When a balance between the number of predators and the prey exists, it is called a *sustainable relationship*.

Can you calculate what happens to a growing prey population (bacteria) when there is also a growing predator (paramecia) population?

Note: There are no live organisms in this project.

MATERIALS

- Calculator (or computer spreadsheet program)
- Graph paper

PROCEDURES

You have an imaginary culture containing 10 paramecia. These paramecia reproduce by fission every 3 hours. Imagine that this species of paramecium eats five bacteria per hour. In this imaginary culture, you also have 100 bacteria. The bacteria reproduce by fission every half-hour. Consider this an ideal setting where the bacteria have an unlimited supply of food.

CONCLUSIONS

How many bacteria and paramecia will you have after 10 hours? After 20 and 30 hours? Is this system sustainable? What would happen if the paramecia ate 10 bacteria per hour instead of five? How does the feeding rate of the paramecia effect the survival of their prey? Slow down the reproductive rate of the paramecia. Can this system now support a higher feeding rate by paramecia?

Graph the results from above (see Fig. 29-1). Put the paramecium and bacteria populations on the y-axis and put the time on the x-axis.

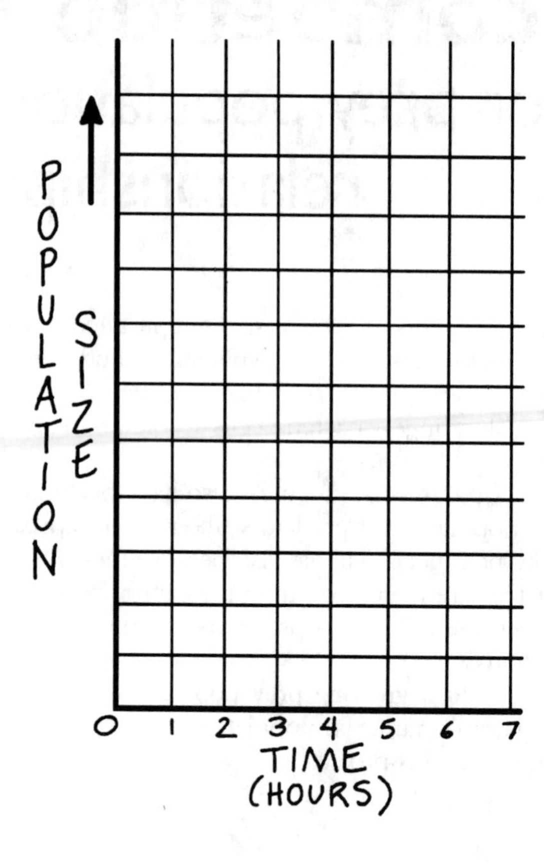

29-1
Plot the predator and prey populations on a graph similar to this one.

GOING FURTHER

Create similar imaginary cultures using other types of microbe predator-prey relationships. Study the population dynamics of a real life predator-prey relationship in a microecosystem.

30

Underwater color

Algae photosynthesis & colored light

Algae are producers, meaning they can create their own food using sunlight during *photosynthesis*. The process of photosynthesis converts carbon dioxide and water into sugar in the presence of light. The sugar is then used as a source of energy.

Visible sunlight is composed of many different wavelengths of light. Different wavelengths of light produce different colors. Does the color of light affect the growth of algae cells?

MATERIALS

- Freshwater pond collection (take surface scum and some plant debris from a pond and place in mayonnaise jar with a lid)
- Five pint-sized mason jars with lids
- Clear, red, green, yellow, and blue cellophane
- 1-inch adhesive tape
- Microscope slides
- Coverslips
- Microscope
- Drawing paper and pencil

PROCEDURES

Take the newly collected freshwater pond sample and divide it equally into five pint-sized mason jars. Put the lids on the jars loosely (to allow air in). Surround each jar with a different color cellophane (see Fig. 30-1). Use the adhesive tape to hold the cellophane in place. Place all the jars in a window that gets indirect sunlight. Don't let them get too warm or all the organisms will die. Leave the jars there for 1 week.

After 1 week, unwrap the cellophane and make observations about the color of the water within the jar. Take a water sample from each jar. Then rewrap the jar and leave it in its original position for another week. Place each sample on a dif-

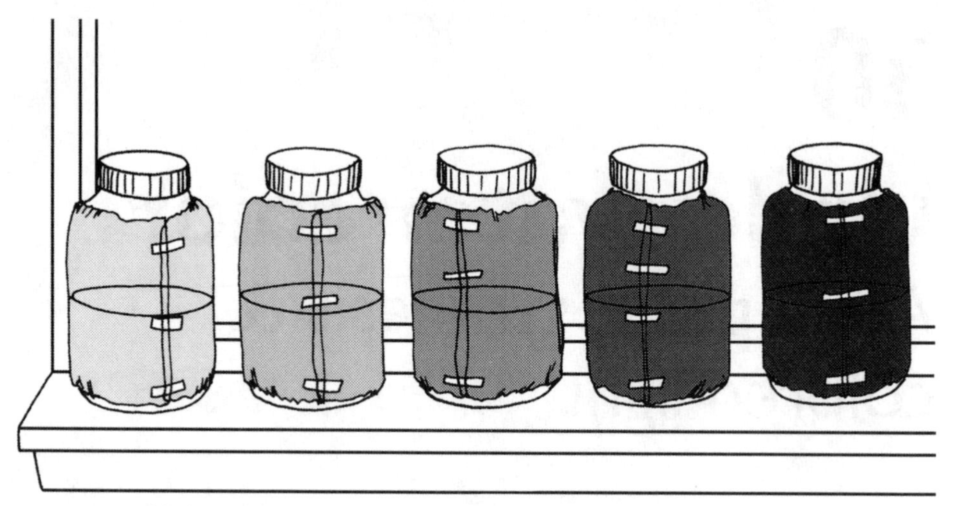

30-1 Wrap each of the jars in a different colored cellophane wrap.

ferent microscope slide. Be sure the slides are marked so you can identify the jar from which the sample came. Place a coverslip on them and view with microscope under 100× and 400× power. Draw the types of organisms that came from each jar. Repeat this procedure after the second week has passed.

CONCLUSIONS

What is the effect of color on algal growth? Do some colors prevent growth, while others encourage it? Are any differences consistent throughout the entire 2-week period or only a portion of it?

GOING FURTHER

Repeat this experiment using a pure culture of algae. How does color affect different species of algae?

31

Miniature snowflakes

Diatoms in different habitats

Diatoms, also known as the golden-brown algae, are surrounded by a shell. Their shells are made of silica and come in an incredible number of shapes and sizes. Diatoms are beautiful works of art, similar in their fine detail to snowflakes, but microscopic in size. Diatoms come in two basic forms: *pennate* (i.e., pen-shaped or elongated) or *centric* (i.e., round or cylindrical).

Diatoms can be found in most freshwater habitats. Are there differences in the types of diatoms found in a freshwater pond versus a freshwater stream? Are there differences between those found in the surface waters and those attached to rocks and plants?

MATERIALS

- Two water collections (about 100 ml each; one from a pond and the other from a stream)
- Filter (0.45 micron) and filter holder (available from scientific supply house)
- Small laboratory spatula
- Microscope slides
- Water
- Forceps and tongs
- Bunsen burner
- Eyedropper
- 5% glycerin (available from a pharmacy or scientific supply house)
- Coverslips
- Drawing paper and pencils
- Two substrate collections (e.g., rocks, pebbles; one from a pond and the other from a stream)

PROCEDURES

You will first take the water sample from the pond and concentrate it to 1 ml by filtering it through a 0.45 micron filter. To do this, use a syringe filter (see Fig. 31-1). Fill the syringe with the water sample and force the liquid through the filter. Use a spatula to scrape off the material that collected on the filter and put it on a microscope slide.

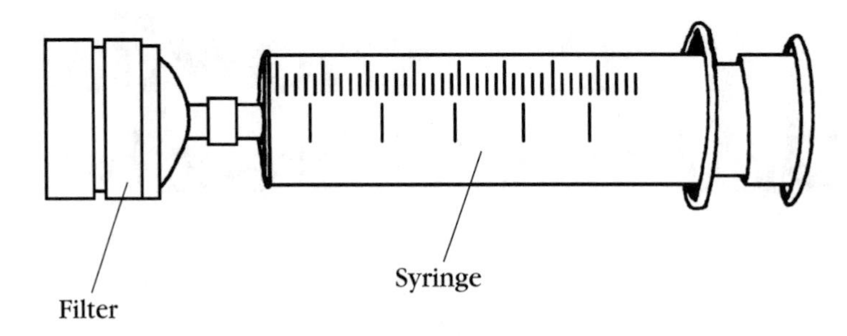

31-1 The sterilization filter is a simple way of removing microbes from a fluid.

Add a small drop of water to the material. Have your teacher or advisor hold the slide with the forceps or tongs over the Bunsen burner until it boils and steams. Remove the slide from the flame and put a drop of 5% glycerin on it. Place a coverslip over the sample and view it under 100× and then 400×. Draw what you see. Repeat this procedure with a sample of water from the running stream or brook.

Next you'll prepare slides from the substrate. Use the spatula to scrape off some scum from the freshwater substrate (e.g., a rock, an underwater plant stem from a freshwater pond). Spread the material onto a slide and place a drop of water over it. Follow the same steps as listed above to heat, steam, and prepare the slide for viewing. Repeat this entire procedure for the substrate from the running stream or brook.

CONCLUSIONS

Did you find diatoms in all four samples? Which had the largest number of different types of diatoms and which had the most diatoms? Compare the stream to the pond and compare the water sample to the substrate sample. What differences do you find? Why do you think some samples had fewer diatoms than others?

GOING FURTHER

Try to identify the diatoms you found. How many different types can you collect from one stream? From one pond? Do this at different seasons. Do you see changes in the diatom community with time?

Part seven

Control of microorganisms

Throughout human history, we have tried to control the growth of microorganisms. The projects in this section explore how we try to protect food and ourselves from harmful microbes. Early methods to preserve foods through the winter were really basic ways of controlling microbes. Sanitation, cleanliness, and removal of human wastes are also methods of microbial control.

The biggest advance in medicine since the discovery that *germs* (microbes that cause disease) existed, was the discovery of substances to kill those germs. *Sulfa* drugs were first used, until the discovery of *antibiotics.* Antibiotics are produced from microorganisms.

Because microorganisms must compete with other microorganisms for food and space, they have developed ways of controlling other microbes around them. Microbes do this by releasing chemicals into the environment that reduce the growth of other microbes near them. These chemicals are the antibiotics that we have learned to obtain and use for our own purposes. Antibiotics have saved millions of lives since their discovery. It is interesting that our best tool to control microbes come from microorganisms themselves.

Chemicals other than antibiotics can also kill or control microbes. Examples include salt, bleach, and soaps. Like all living things, microbes must be in the correct type of environment for them to live. If the chemicals in their environment are wrong, they can't grow or survive. Humans have learned to adjust the chemicals in their environment to control microbe growth.

Chemicals used to control microbes include *phenol* (found in some mouthwashes) and *chlorine* (found in bleach and used to disinfect drinking water and swimming pools). Iodine, alcohols, hydrogen peroxides, and detergents are other examples used to control microbes. It is important to realize that some of these substances kill microbes (called a *germicide*), while others only prevent their growth (called a *germistat*).

Physical factors such as heat, temperature, sound, and light are important to all forms of life. Most organisms have an ideal range of temperature and moisture. If they exist out of that range, they can't survive. These factors can be used to control microbes. No organisms can survive the heat from a flame, therefore inoculating loops are sterilized by flame. Intense sound, such as ultrasonic waves, disrupt the cells of microbes, and can be used in some situations. Electromagnetic radiation, such as X-rays, will kill microbes. Cold, such as in a freezer, doesn't always kill microbes, but stops their growth.

32

Help for the wounded

Comparison of over-the-counter antibiotics

Anytime your skin is wounded, it is open to the entry of microbes. Skin is our front line of defense against microbes. Without it, our bodies would have to constantly fight infections (i.e., growths of foreign microbes). When our skin is broken, it is no longer able to defend against the entry of microbes.

When we get cuts or scrapes, we often apply medications available at the drugstore. Compare over-the-counter medications people use to treat wounds or abrasions. Do some protect us better than others? Do some only protect us from certain types of microbes?

MATERIALS

- Six sterile nutrient agar plates (available from scientific supply house)
- Marker (that writes on plastic)
- Three sterile cotton swabs
- A known nonpathenogenic gram-positive culture (e.g., *Bacillus subtilis, Bacillus cereus, Clostridium botricum, Streptococcus lactis;* available from scientific supply house)
- A known nonpathenogenic gram-negative culture (e.g., *Rhizobium leguminosarum, Spirillum volutans, Vibrio fischeri;* available from scientific supply house)
- A mold culture (e.g., *Neurospora crassa, Saccharomyces cerevisiae;* available from scientific supply house)
- Bacitracin antibiotic cream (available from a pharmacy or drugstore)
- A first-aid cream
- A triple-antibiotic cream

- Tenactin cream
- 1-inch adhesive tape
- Thermometer

PROCEDURES

Label each of the sterile nutrient agar plates with one of the following: "Gram+," "Gram+/Control," "Gram–," "Gram–/Control." Dip one sterile cotton swab into the gram + culture (if the culture is in solid form, such as an agar slant or a plate, roll the swab over the surface of the culture a number of times). Remove the swab and roll it over the plate labeled "Gram+" to give the plate an even coat of microbes (see Fig. 32-1). Repeat this same procedure for the plate labeled "gram+/control." Dispose of the cotton swab properly.

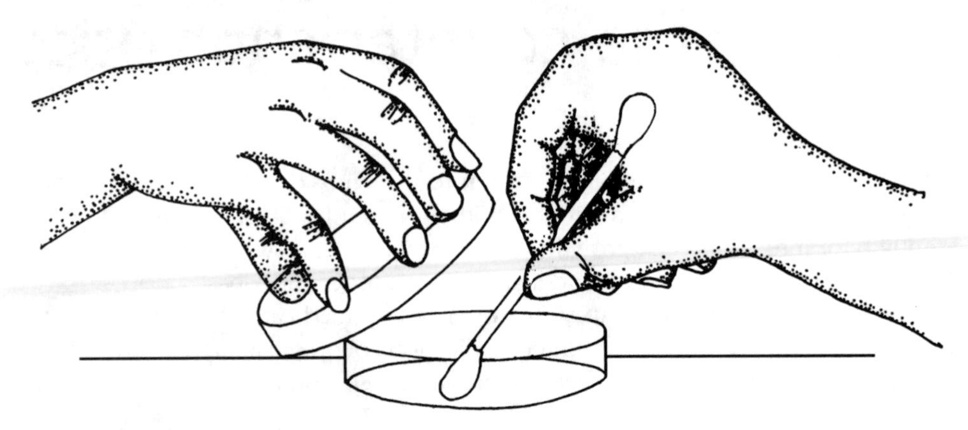

32-1 While holding the cover up with one hand, gently roll the swab over the nutrient agar.

Repeat this procedure for the two plates labeled as "gram–." Once all the plates are inoculated, draw lines over the lids on the plates that are *not* labeled "control" to divide the plates into quarters (see Fig. 32-2). Label each of the quarters in each of these plates with one of the following: "Bacitracin," "First-aid," "Triple," and "Tenactin." Place a drop (¼ inch diameter) of each cream in the center of each quartered section in the plate. The two plates labeled as controls will not receive any antibiotics. Finally, seal all the plates with adhesive tape.

Incubate the plates at 37 degrees C, preferably under a lamp. Try to keep the temperature between 30 to 35 degrees C. After 2 days, observe all the plates for growth. If the control plates are not growing, or if they are contaminated, repeat the experiment, using good sterile technique (see chapter 1 "An Introduction to Microbiology" on the use of good sterile techniques).

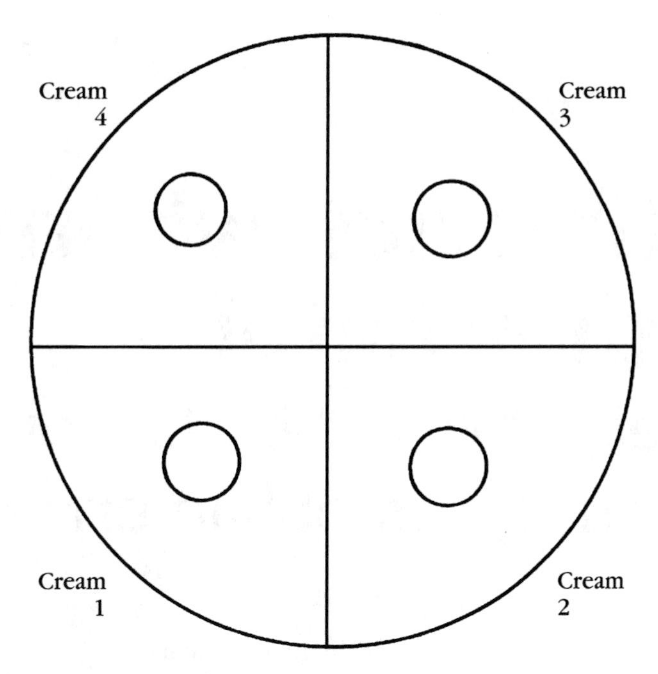

32-2 Place a drop of antibiotic cream in the center of each quadrant of the plate.

CONCLUSIONS

Do these over-the-counter medications control nonpathenogenic microbe growth? Do different medications control different types of microbes (e.g., gram+ and gram–)? Are there different microbial growth patterns around the drops of wound creams? Are there zones where no growth is occurring? What do you conclude about the effects of these over-the-counter creams and their effects on different types of microbes?

GOING FURTHER

Try using other types of microbes and other over-the-counter antimicrobial agents. Study the history of the development of antibiotics.

33

Antibiotics–whom do they hate?

The effect of antibiotics on different types of bacteria

There are dozens of different types of antibiotics that a doctor may prescribe for a patient with an infection. These tools of medicine have made life for humankind much easier and safer. *Antibiotics* are chemicals that are produced by microorganisms to kill or inhibit the growth of other microorganisms. Scientists have learned to collect these chemicals and use them to enhance or inhibit the growth of microbes in humans. Now scientists can create a few of our own antibiotics to control some microbes.

A common way to test how susceptible a microbe is to an antibiotic is by the *disk diffusion test.* The antibiotic is placed on a disk of paper that is then placed on top of a microbe culture. If the culture continues to grow under and around this disk, the culture is *not* susceptible to the antibiotic. If the culture does *not* grow around the disk, it is susceptible to the antibiotic. Are different classification groups of bacteria susceptible to one type of antibiotic? For example, are gram negative and gram positive bacteria susceptible to the same antibiotics?

MATERIALS

- Marker that writes on plastic
- Four sterile nutrient agar plates (available from scientific supply house)
- Two sterile cotton swabs (available from a drugstore, medical supply store, or a scientific supply house)
- Known gram-positive culture (e.g., *Bacillus subtilis, Bacillus cereus, Clostridium botricum, Streptococcus lactis;* available from scientific supply house)
- Known gram-negative culture (e.g., *Rhizobium leguminosarum, Spirillum volutans, Vibrio fischeri;* available from scientific supply house)

- Four different types of antibiotic disks (available from scientific supply house)
- 1-inch adhesive tape
- Incubator (or warm area in room under lamp)

PROCEDURES

Label the bottom of each plate with one of the following: "Gram+," "Gram–," "Gram+/Control," and "Gram–/Control." Dip a sterile cotton swab into the gram positive culture. If the culture is a solid culture, roll the swab over the surface of the culture a number of times. Roll the swab over the plates labeled "gram+" to give the plate an even coat of microbes. Take a clean swab and dip it into the gram negative culture. Again, if the culture is a solid culture, roll the swab over the surface of the culture a number of times. Roll the swab over the plates labeled "Gram–" to give the plate an even coat of microbes.

Draw lines on the lids of both of the plates that are not labeled as "control" to divide the plates into quarters. Label all the quarters with the name of each of the four types of antibiotic disks. The names will be provided with the antibiotic disks. Put the proper disk in the center of the matching quartered space on the plate. Seal all the plates with adhesive tape.

Incubate the plates at 37 degrees C (or under a lamp to keep the temperature between 30 and 35 degrees C). After 2 days, look at all the plates for growth (see Fig. 33-1).

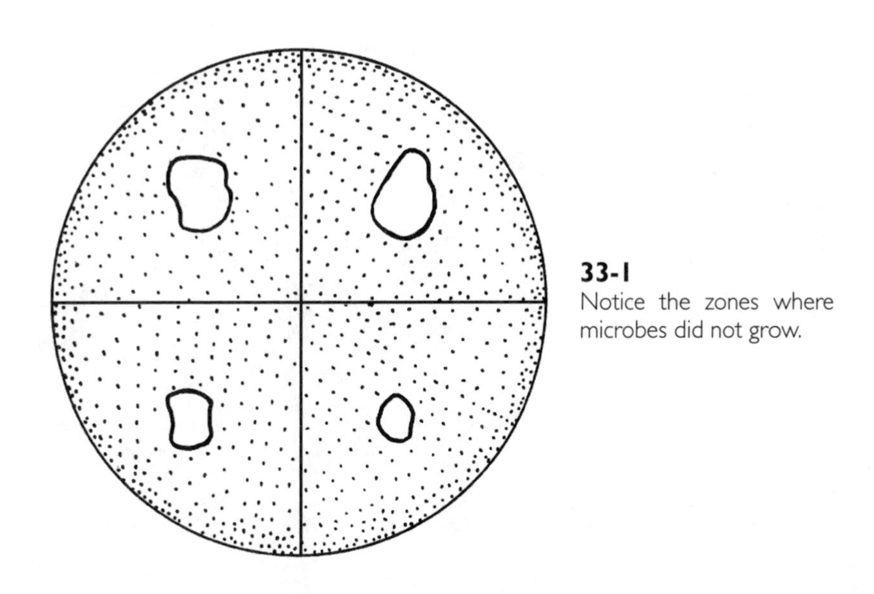

33-1
Notice the zones where microbes did not grow.

CONCLUSIONS

Are there different microbial growth patterns around the four disks? Are there zones where no growth is occurring? What do you conclude about the effects of these four antibiotics on gram-negative and gram-positive bacteria? Are there differences in susceptibility between gram-positive and gram-negative types of bacteria?

GOING FURTHER

Test this with other microbes and antibiotics. Study the history of the research and development of antibiotics.

34

Spray them till they drop

Household antimicrobial sprays

Household disinfectants and antimicrobial sprays are used to reduce or eliminate microorganisms that live with us. They are especially important in hospitals. Because hospitals are the temporary home of sick people, there are dangerous microbes concentrated in hospitals. Disinfectants cut down on the microbes living on surfaces in our homes, in hospitals, and in other buildings.

Surfaces in our homes also contain microbes, some of which are disease-causing, but most of which are not. Bathrooms are a favorite room for microbes. Many people use household disinfectants to clean their bathrooms. How good are common household disinfectants at reducing microbes?

MATERIALS

- Marker (that writes on plastic)
- Four sterile nutrient agar plates (available from scientific supply house)
- A toilet fixture (you will only be using the base of the fixture, near the floor)
- Rubber gloves
- Sterile bandages
- Household spray disinfectant (e.g., Lysol)
- Homemade disinfectant (e.g., 1 part bleach to 100 parts water)
- Soapy water
- 1-inch adhesive tape
- Incubator (or warm area in room under lamp)
- Thermometer

PROCEDURES

Label each of the plates with one of the following: "None," "Wipe," "Spray Disinfectant," "Homemade Disinfectant," and "Soapy Water." Imagine that the *base* of the toilet is divided into four sections (see Fig. 34-1). Each section will receive a different disinfectant treatment that will be used to inoculate (transfer) microbes to the plates.

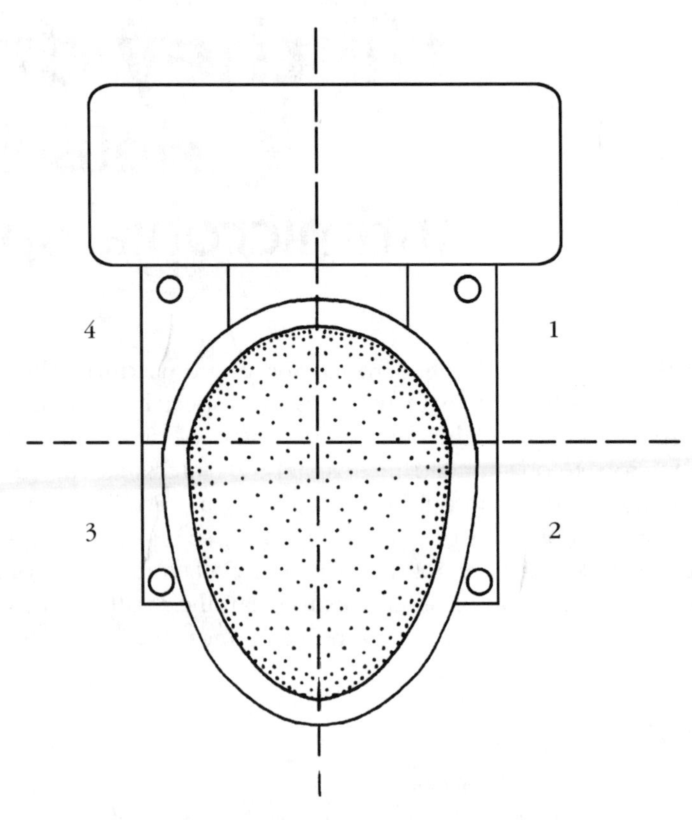

34-1 Imagine the base of the toilet bowl divided into four sections.

Put on a pair of very clean, dry rubber gloves. Over the gloves, put a bandage on your index finger backwards as you see it (see Fig. 34-2). Touch the sterile gauze portion of the bandage to the bottom of the toilet (where the toilet meets the floor) in section "one." Keep the lid to the bowl closed. *Do not place your hands in the bowl at any time.* Then touch the gauze to the plate labeled "None." Discard the bandage, and put another backwards bandage on your finger as before. Once again, touch the gauze to the same section of the toilet and then touch the gauze to another area on the same "None" plate. Discard that bandage. Finally, circle on the lid of the plate, the two regions you touched with the contaminated bandage (see Fig. 34-3). The "None" plate is now complete.

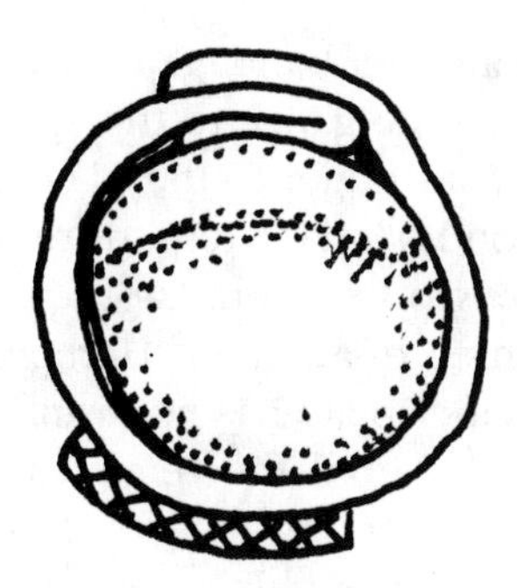

34-2 Wrap the bandage (backwards) around your gloved finger as shown.

34-3 Circle the two spots where you touched the gauze to the plate.

Apply one of the disinfectants (e.g., spray disinfectant, homemade disinfectant, soapy water) to each of the other three sections of the toilet. Wait for 2 minutes before proceeding. With different, clean, paper towels, gently wipe off any excess liquid that might remain on the toilet. Also wipe the "None" quadrant (this will be a test to see if any microbes are removed by the wiping motion rather than by the action of the disinfectant).

While wearing the rubber gloves, put another bandage on your index finger, backwards, as you did earlier. Touch the sterile gauze to the section where the spray disinfectant was used. Then touch your finger to the plate labeled "Spray Disinfectant" as you did earlier. Use a second bandage to perform the second inoculation in the same plate and circle the two fingerprints on the lid.

Repeat this process with the other three quadrants: "Homemade Disinfectant," "Soapy Water," and "Wipe." (The "Wipe" plate has no disinfectant.) Seal all the plates with adhesive tape. You should end up with five plates, each with two circles indicating where you touched your finger to the plate. Be sure to dispose of the gloves properly and wash your hands throughly when done. Incubate all the plates in an incubator at 37 degrees C or in a warm area in a room under a lamp. Note the temperature with a thermometer and be sure the temperature doesn't fluctuate more than 2 degrees. Observe growth patterns twice a day on each plate.

CONCLUSIONS

Do you get growth on all the plates with all the disinfectants? How long did it take for the growth to start? Did the growth start at the same time in all the plates? What can you conclude about the action of these disinfectants in preventing microbial growth?

GOING FURTHER

Use different disinfectants. Are they all as efficient or inefficient? Do this in different areas of the house. Do you get different results?

35

Sugar & salt

Preserving meats

Humans have always competed with microbes for food. Humans struggle to keep microbes out of their foods so they do not spoil or become poisonous. Before refrigeration and canning, it was crucial that enough food be preserved for the winter months. Chemicals, such as salt and sugar, were used to control microbes in our foods. Does salt and sugar prevent or control meat from spoiling?

MATERIALS

- Marker that writes on plastic
- Five resealable plastic bags (Ziploc-type)
- Five small pieces of thin (less than ¼-inch thick) round steak
- Sink
- ½ teaspoon measurer
- Salt
- Broiler pan
- Towel
- ½ tablespoon measurer
- Sugar

PROCEDURES

Label each of the five resealable plastic bags with one of the following: "0," "1 Teaspoon Salt," "1 Tablespoon Salt," "1 Teaspoon Sugar," and "1 Tablespoon Sugar." Lay a slice of meat in a dry, clean sink. Sprinkle ½ teaspoon of salt onto meat. Try to distribute the salt evenly by rubbing it in all over the slice. Turn the slice over, and rub in a second ½ teaspoon of salt. Put the slice on the broiler pan. The broiler pan will soak up any liquids that run off. Put the "1 Teaspoon" labeled bag next to this slice (this will mark which slice has what treatment). Wash out the sink and towel it dry.

Place a second slice of meat in the sink, and rub ½ tablespoon of salt onto one side, distributing it evenly. Turn the slice over and rub in another ½ tablespoon of salt. Put this slice on the broiler pan and mark its position with the "1

Tablespoon Salt" bag. Clean and dry the sink once again, and repeat the entire procedure for the 1 teaspoon of sugar slice and the 1 tablespoon of sugar slice. Finally, put the slice of meat without sugar or salt on the broiler pan (see Fig. 35-1). Let the pieces sit for 4 hours, then place each slice in its bag and seal the bags.

Check all the slices every 12 hours, for 14 days for microbial growth. You can either observe the growth from the outside of the bags, or you can open the

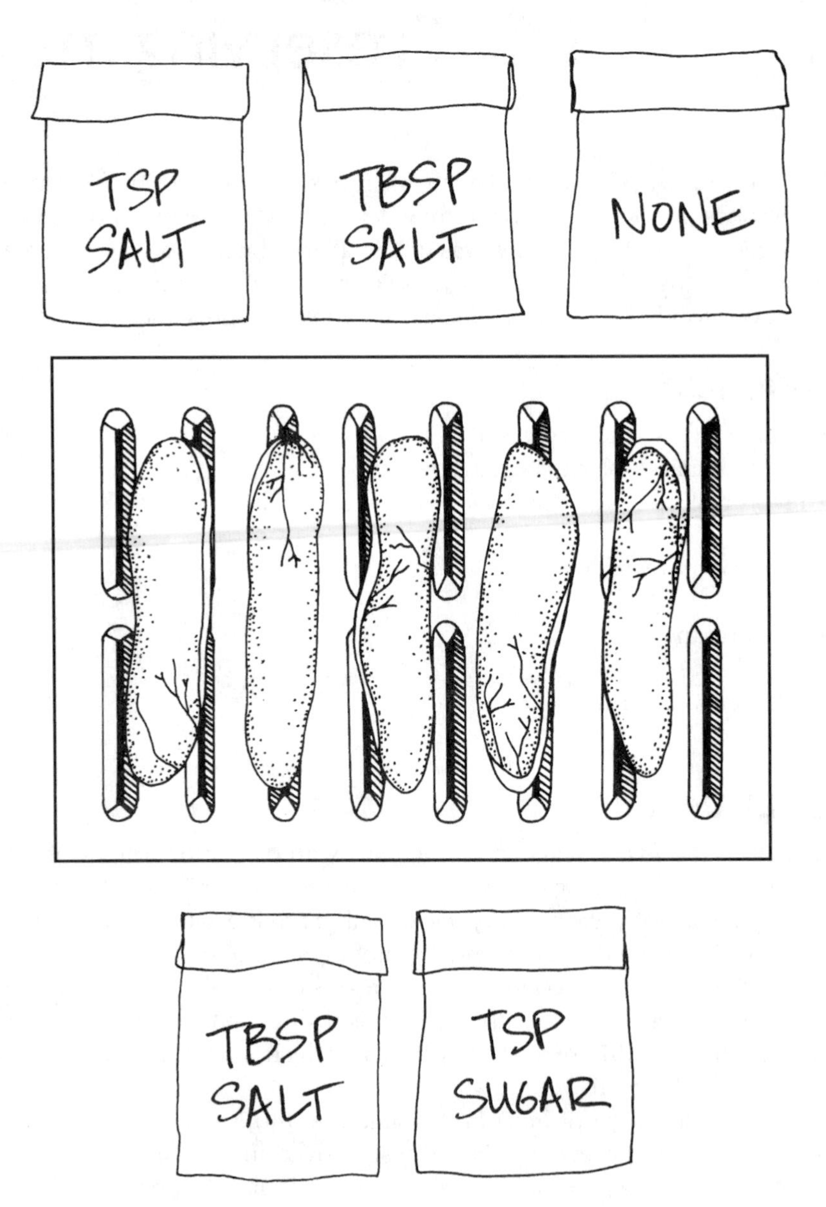

35-1 Place each labeled bag next to the seasoned meat.

bags and check the odor of the meat. Note when the odor is that of spoiled meat and/or when the slice appears slimy or moldy. Note the texture of each slice over time. Dispose of any piece of meat that has an excessive smell, but be sure to note the condition and date on which it was disposed.

CONCLUSIONS

Do all the slices show growth and begin to smell at the same time? Does salt or sugar act as a preservative? Do different amounts of the substances used appear to affect the spoilage? What do you conclude about salt and sugar as meat preservatives?

GOING FURTHER

Investigate why some substances preserve meats and why others don't. Study how some of these substances were used long ago to preserve foods. Research the relationship between these preservation methods and what it means for meat to be considered kosher.

36

Falling apart at the seams

Denaturing common proteins

Temperature plays an important role in a microbe's life. Too little heat slows the speed of chemical reactions (*metabolism*), and the microbe can't grow. When there is too much heat, the proteins (what most cells are made of) change their shape and lose their ability to function. When proteins lose their natural shape, they are said to have *denatured*. This relationship between heat and protein denaturing is used to kill microbes. Examples of denaturing include using a flame to sterilize an inoculating loop or pasteurization to kill microbes in milk.

There are many common proteins we can use to study the process of denaturing. Egg whites are made primarily of a protein called *albumin*, *casein* is the major milk protein, and your hair is composed mostly of a protein called *keratin*. Do these different proteins have different temperatures at which they denature?

MATERIALS

- Saucepan
- Egg whites
- Candy thermometer
- Stove
- Spoon
- Reconstituted nonfat dry milk
- Hair (from an individual's hairbrush)
- Drawing paper and pencil
- Cookie sheet
- Aluminum foil
- Oven
- Oven thermometer
- Oven mitt

PROCEDURES

Fill the bottom of the saucepan with egg whites. Put the candy thermometer in the saucepan. Put the saucepan on the stove at medium heat. Occasionally stir the contents of the pan with the spoon. At what temperature do the egg whites begin to change their texture and consistency, meaning the protein has denatured? Clean and dry both the pan and the spoon. Repeat this procedure with the reconstituted nonfat dry milk. Note the temperature that the milk begins to change consistency, meaning the protein is becoming denatured. Watch the pan closely so that the milk does not boil over.

Take a small clump of hair (less than a thimble full) from a hairbrush. Note the texture and shape of the individual hairs. You might want to draw a picture of them. Use hairs from the same person, since hair texture is not always the same between individuals. Put the clump of hair in the middle of a cookie sheet, on a piece of aluminum foil. Put the oven thermometer in the oven, and place the cookie sheet in the oven. Slowly increase the oven temperature up to 200 degrees F for 30 minutes. Increase the oven temperature by 50 degree F intervals until the hair texture begins to change. Use an oven mitt to remove the cookie sheet.

CONCLUSIONS

What can you conclude about the temperature required to denature the proteins albumin, casein, and keratin? Do they all denature at about the same temperature?

GOING FURTHER

Compare how the proteins in microbes denature as compared to those in the proteins mentioned above.

37

Hitchhiking microbes

Insects transporting microbes

When trying to control microbes, it is important to understand where and how they get around. Air and water can transport microbes, but they can travel in other ways too. Other organisms can transport microbes from one place to another. Can organisms, such as insects, transport microbes from one place to another?

MATERIALS

- One sterile nutrient agar plate (available from scientific supply house)
- One live ant (or other insects)
- Marker (that writes on plastic)
- Forceps
- 1-inch adhesive tape
- Magnifying glass or stereoscope

PROCEDURES

Place a live ant on the nutrient agar plate. Close the plate. Let the ant walk around on the agar for 1 minute. As it walks around in the plate, use a marker to draw the trail on the lid of the plate that follows the ant's actual trail (see Fig. 37-1). Then, turn the plate over to dump out the ant to release or dispose of it.

Seal the plate with adhesive tape and incubate it in a warm area of the room at 30 degrees C for 24 to 48 hours. After that time, observe the plate for microbial growth. Use a magnifying glass or stereoscope to see the colonies in more detail.

37-1
Draw the trail of the ant on the top of the plate lid.

CONCLUSIONS

Have the colonies of microbes grown along the trail where the ant walked? Draw the growth patterns. What do you conclude about an ant's ability to transport microbes?

GOING FURTHER

Use different types of insects, such as a cockroach that usually lives indoors and a cricket that usually lives outdoors. Observe the microbes that grow under a scope and try to identify them.

38

How clean is clean?

Washing utensils

Microbes are all around us, including on the utensils that we think are clean. Bacteria and other microbes settle on utensils and can be transferred to an environment where they can grow. When we clean our hands or wash our dishes, most people think that we are removing all the germs. Is this true? Are there microbes found on our eating utensils before and after cleaning? Does cleaning with a dishwasher make it any cleaner?

MATERIALS

- Six sterile nutrient agar plates (available from scientific supply house)
- Marker (that writes on plastic)
- Two dirty forks that had been used for eating a meal within the last 1 hour
- Two clean forks that have been washed by hand and air-dried after being used for a meal
- Two forks that have been washed and dried in a dishwasher after being used for a meal (optional)
- Incubator (or warm area in the room under lamp)
- Thermometer

PROCEDURES

Label two of the sterile nutrient agar plates "Clean/Handwash," the second two "Clean/Dishwash," and the last two "dirty." Take one of the "Dirty" forks, and press the tines into the agar surface of a plate labeled "Dirty" (see Fig. 38-1). Repeat this with the other "Dirty" fork using another plate. Repeat the same procedure for all four clean forks. You'll end up with three sets of two plates.

Place all the plates in an incubator, if available, at 37 degrees C or keep in a warm area in the room. Note the temperature of the area and check it with a thermometer over time to be sure it doesn't fluctuate more than 2 degrees. Incubate the plates for 48 hours. Observe and make notes of the number, placement, and shape of colonies on the plates.

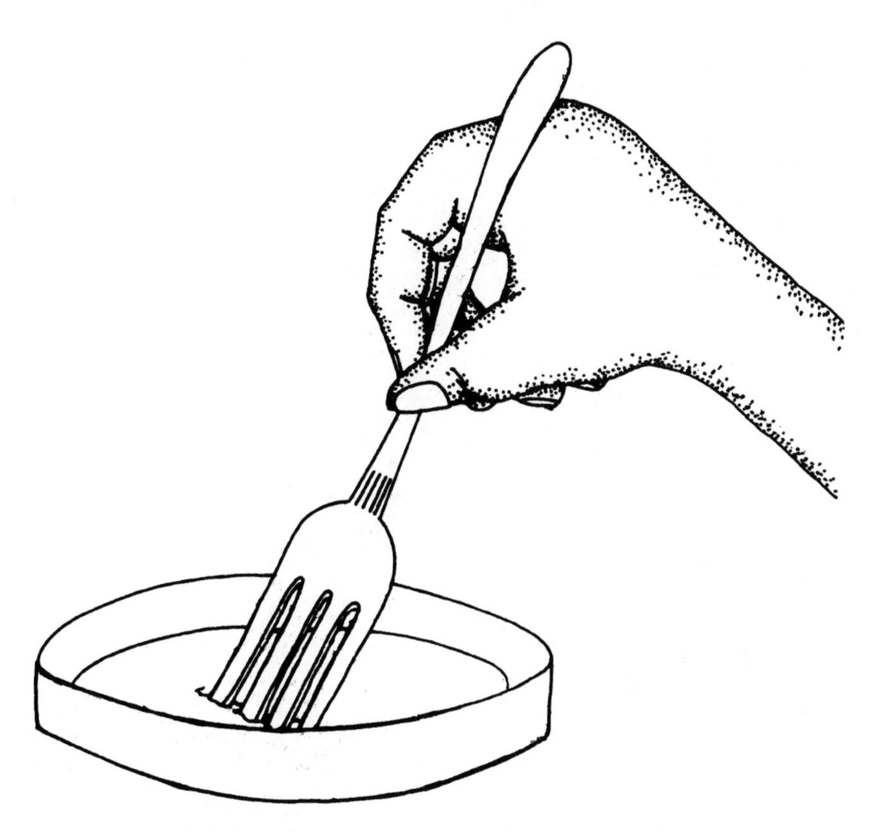

38-1 Gently press the fork tines into the agar plate.

CONCLUSIONS

Are there differences between the clean forks and the dirty forks? Are there differences between the two types of clean forks? Do the "clean" plates have any growth? What can you conclude about bacteria on the surfaces of so-called clean utensils? Does a dishwasher make a difference in the number of microbes found on the fork?

GOING FURTHER

Repeat this project with forks cleaned in the dishwasher using different types of dish detergents. Are there any differences? Also, compare the growth found from a clean fork over different periods of time since it was cleaned. How rapidly do microbes accumulate on the forks?

CONCLUSION

Part eight

Microbes & disease

Microbes that cause disease are called *pathogens.* Diseases caused by pathogens used to be the biggest killer of humans. The study of microbiology has led to many ways of controlling and curing diseases caused by pathogens.

Some microbes cause disease in humans by accident. These organisms don't normally live on or in humans, but if they do, they can make us sick. Good sanitation and cleanliness reduce the possibility of this happening. Other disease-causing microbes always live in the body of another organism, called the host. These organisms are harder to control and protect ourselves against.

The term *infection* refers to the growth of a microorganism in a host. Not all infections lead to disease. In many cases, we can fight off an infection or in some cases, the infection causes no harm and we simply coexist with these microbes.

Pathogens are transmitted from one host to another in many ways. Some are airborne—they can find a new host simply by being inhaled. Cold viruses often get to their new host by contact. For example, being picked up on a noncontaminated hand after the contaminated hand was in contact with an infected nose or eyes.

There are three steps to disease transmission. First, the pathogen must escape from its original host. Coughs, mucus secretions, or simply breathing could be the means of escape. Second, the pathogen must travel outside of a host's body to reach a new host. And last, the pathogen must enter the body of a new host. If any one of these steps fail, then transmission does not occur. The projects in this section explore the identification, diagnosis, and transmission of pathogens.

39 The sick little bug
Microbial disease in insects

Many microbes are *parasitic*; that is, they need another organism to survive. Some parasites don't hurt their *hosts*, the organisms in which they live; others cause illness, disease, and even death to their host. The correct conditions need to be present for the parasitic microbe to survive on the host.

Many human diseases (e.g., measles) are caused by microbial parasites, but humans aren't the only animal affected by these small creatures. Numerous other types of organisms can be host to these parasites. There are even parasites on insects. What environmental conditions (e.g., temperature, moisture) are needed for *Beauvaria brassicae*, a fungal parasite of insects, to infect and kill crickets?

MATERIALS

- Scissors
- Four empty cottage cheese containers (16 oz.) with lids
- Newspaper
- Fresh apple slices with skin
- Crickets (available from pet store, supply house, or you can collect your own)
- A freeze-dried culture of *Beauvaria brassicae* (available from scientific supply house)
- Marker (that writes on plastic)
- Water sprayer/mister
- Water
- Cool area in building
- Warm area in building
- Thermometer

PROCEDURES

Use the scissors to punch four air holes (less than ¼ inch in diameter) in the cottage cheese container lids. Put a layer of newspaper on the bottom of each container. Cut the paper to fit so it can lay flat. Put one small apple slide, less than 1/16 of apple, in each container and add two crickets to each container (see Fig. 39-1).

39-1 Put newspapers in the bottom of the container, and then add one slice of apple and two crickets.

Inoculate the containers with *Beauvaria brassicae* by gently tapping the vial containing the freeze-dried *Beauvaria brassicae* into each cricket container. Put the same amount of the dried fungus into each cricket container. Put the lids on all the containers. Label the containers as follows: "Cool/Dry," "Cool/Moist," "Warm/Dry," and "Warm/Moist."

Every day, use the water sprayer to lightly mist the interior of the two containers labeled "moist." Place the two "Cool" containers in a cool area of the building, such as the basement, and put the two "warm" containers in a warm area under a lamp. Be sure the temperature under the lamp doesn't go above 28 degrees C. Every 2 to 3 days, replace the apple slice with a fresh slice.

Each day, check the containers. If you can't find the crickets, check under the newspaper. If a cricket dies, does its body change appearance? Leave the dead cricket in its container and continue daily observations. Note the time of any external changes to the cricket's body. For example, when does it appear moldy?

CONCLUSIONS

Do certain conditions make crickets more prone to being attacked and killed by microbes? What can you conclude about the effect of temperature and humidity on the growth of *Beauvaria brassicae* on crickets?

GOING FURTHER

What kind of microbe is *Beauvaria brassicae*? Can other insects be infected with *Beauvaria brassicae*? Research if this microbe causes problems for crickets or other insects in the wild. Investigate how these and other microbes are used to control harmful insects.

Gums

40

Cell breakers

ding bacteria on your gums that break cells

luce *enzymes*, which are proteins that help biochemical reactions on, reproduction, and growth all occur because of different en- bacteria, especially *pathogenic* (disease-causing) bacteria, secrete enzymes outside of their bodies (i.e., *exoenzymes*). These exoenzymes can break down the tissues of a bacteria's host. Some of these exoenzymes cause blood cells to break open. This kills the blood cells and helps the bacteria use the nutrients found within these cells. Many disease-causing bacteria have this characteristic.

Can you find any bacteria on your body that use exoenzymes to break open blood cells (i.e., *hemolytic bacteria*)?

MATERIALS

- Sterile cotton swabs (available from pharmacy or a drug store)
- Blood agar plates (available from scientific supply house)
- Marker that writes on plastic
- 1-inch adhesive tape
- Incubator (or warm area in room under lamp)
- Thermometer
- Plastic bags

PROCEDURES

Rub a sterile cotton swab all around your gums, at least 4 hours after you last brushed your teeth (see Fig. 40-1). Take the swab sample and gently roll it over the surface of blood agar plate. Close the plate and label the bottom of this plate "Gum." Next, take another sterile swab and rub it over your skin. Use this swab to inoculate the plate labeled "Skin," then close and label it.

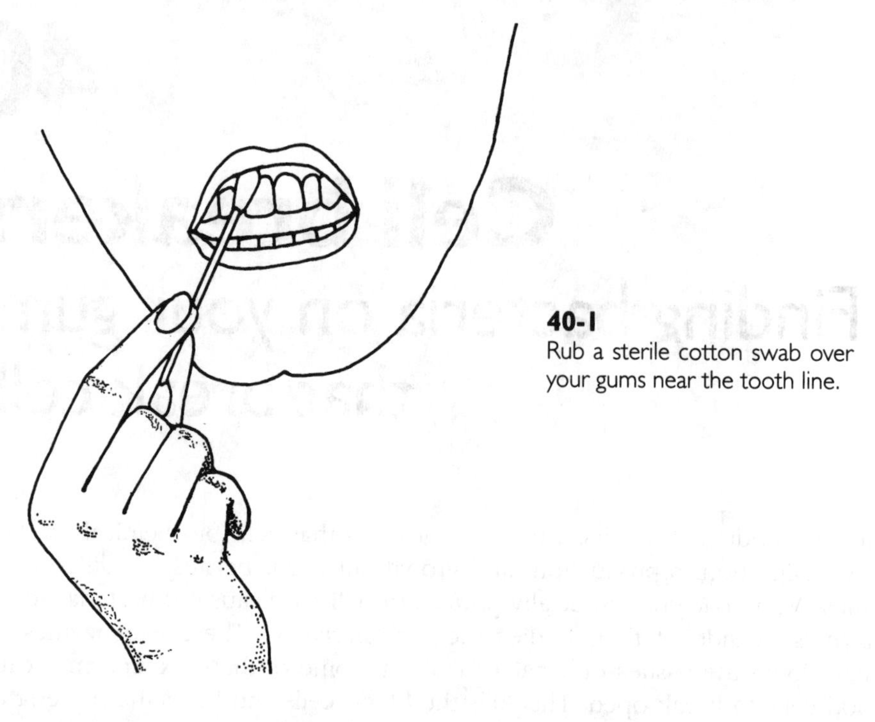

40-1
Rub a sterile cotton swab over your gums near the tooth line.

Next, repeat this procedure by swabbing the back of your throat with another sterile swab and create a "Throat" plate. Finally, take another plate, open it, and cough directly onto it. Immediately close the plate and label it "Cough." Seal all the plates by taping adhesive tape along their edges.

Place the plates in an incubator at 37 degrees C or in a warm area in the room under a lamp. Note the temperature of the area with a thermometer and check it over time to be sure it doesn't fluctuate more than 2 degrees. Watch the plates each day and note the microbial growth. Each day, note the shape of the colonies, their color, and the color of the blood agar media. If the blood cells in the media break (*lyse*), they will lose their red color.

When you are done with the plates, carefully dispose of them by sealing them into a plastic bag, and dispose of them according to your teacher's or advisor's instructions.

CONCLUSIONS

Did you find any hemolytic bacteria in any of the samples? What part of your body contains these bacteria?

GOING FURTHER

Read about hemolytic bacteria. How do our bodies protect against them? How are hemolytic bacteria involved in gum disease? Swab your gums before and after brushing your teeth—does this affect hemolytic bacterial growth? Try the same experiment, but use different types of mouthwash first.

41

Cigars, spheres, & spirals

Identifying bacteria by their shapes

One of the earliest and most basic ways of identifying bacteria is their shape. Single bacteria come in three forms: rods (i.e., cigar-shaped), cocci (i.e., sphere-shaped), and spirilla (i.e., spiral-shaped) (see Fig. 41-1). Some bacteria also grow in groups that form chains of rods or cocci. What are the shapes of microbes living on your gums?

MATERIALS

- Microscope with oil-immersion objective
- Sterile cotton swabs (available from drug store, pharmacy, or scientific supply house)
- Four sterile nutrient agar plates (available from scientific supply house)
- India ink
- Eyedropper
- Microscope slides
- Sterile inoculating loops
- A known rod-shaped bacteria culture (e.g., *Bacillus cereus*; available from scientific supply house)
- Immersion oil
- Coverslips
- Drawing paper and pencils
- A known coccus-shaped bacteria culture (e.g., *Staphylococcus saprophyticus*; available from scientific supply house)
- A known spirillum-shaped bacteria culture (e.g., *Spirillum volutans*; available from scientific supply house)

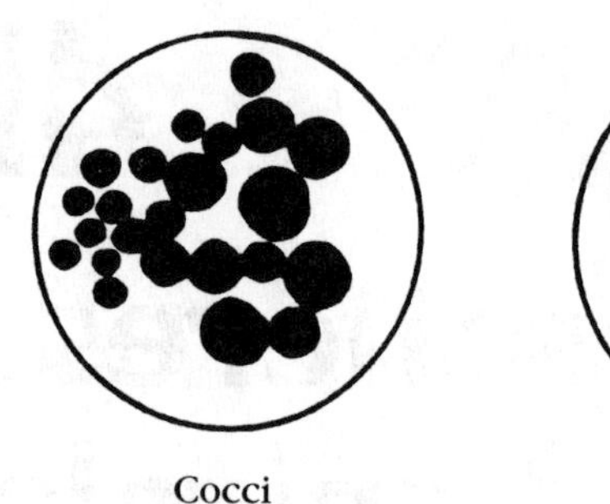

41-1 Bacteria come in three basic shapes, as seen here.

PROCEDURES

Learn how to use the oil-immersion objective of the microscope (see chapter 1 "An Introduction to Microbiology" for instructions about using oil immersion). To create a culture of microbes from your gums, run the sterile swab along your gums near the tooth line. Use this swab to then inoculate a nutrient agar plate by gently rubbing the swab over the agar. Let the plate incubate for 24 to 48 hours in a warm area in the room under a lamp. While waiting for this culture to grow, proceed with the other part of the experiment below.

Put a drop of India ink on a clean glass slide. Use a sterile inoculating loop (see chapter 1 "An Introduction to Microbiology" for instructions about how to sterilize and use an inoculating loop) to take a loop full of the known rod-shaped culture. Mix it into the drop of India ink and spread the suspension around the slide (see Fig. 41-2). Let the slide air dry. Now view it under the microscope at 100× and 400×.

Put a drop of immersion oil on the center of the coverslip and view with the oil immersion objective (1000×). You might be able to see some detail under 400× if the oil-immersion objective is not available. The India ink is not absorbed by the microbes, but it surrounds them which lets you see their shapes better. Draw their shapes.

Repeat this same procedure with the known coccus-shaped culture and the known spirillum-shaped culture. Finally, repeat this procedure with the unknown culture taken from your gums.

CONCLUSIONS

Can you distinguish the different bacterial shapes in the known cultures? Do any of the four cultures grow in linked chains? How do the pure cultures differ from your gum culture? Did you find one shape or different shapes from the gum culture?

GOING FURTHER

Take swab samples from other parts of your body and culture them. Do different parts of your body contain different types of bacteria?

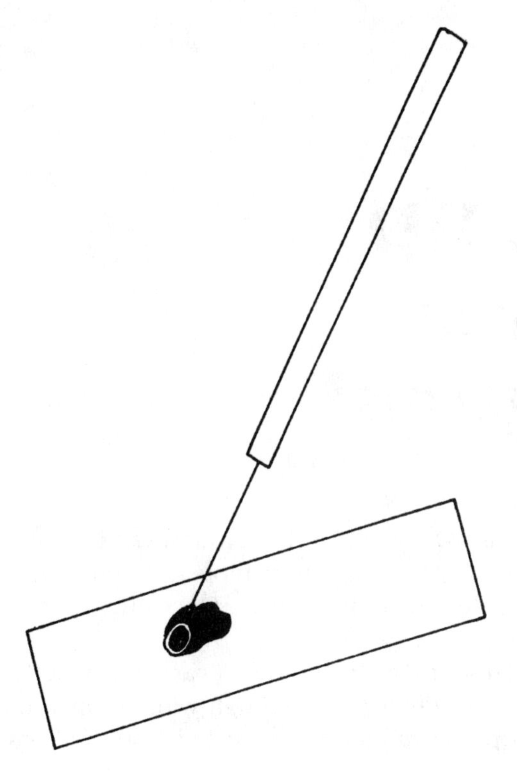

41-2
Move the inoculating loop containing the culture back and forth through the drop of India ink.

42

Cough it up

How far do you cough your germs?

Pathogens (disease-causing organisms) are spread from a host (an infected person) to another person. Air is one way to do it, but pathogens often need to be in a droplet of water to be transferred and to grow. Once they dry, many pathogens will die.

Because water droplets are denser than air, air doesn't carry them too far because the droplets fall out of the air. When people cough with an open mouth (or sneeze), they expel water droplets at a high rate of speed. The speed carries the droplets for some distance. How far can your cough carry the microbes from your mouth?

MATERIALS

- Marker (that writes on plastic)
- Eleven sterile nutrient agar plates (available from scientific supply house)
- Meter or yard stick
- Bench or table that is at least waist-high and 6 feet long
- Two people to assist you
- Surgical face masks (available from a medical supply store, scientific supply house, or a doctor you know)
- 1-inch adhesive tape

PROCEDURES

You need to do this project in a room with very little air movement. Label the plates as follows: "½," "1," "1½," "2," "2½," "3," "4," "5," "6," and "Control." You need a 6-foot-long lab bench or table and another table a few feet behind it (see Fig. 42-1). Place the closed agar plates at each of the distances mentioned above (e.g., ½ foot, 1 foot) and place a 10th plate (i.e., the "Control" plate) on the table behind the long one.

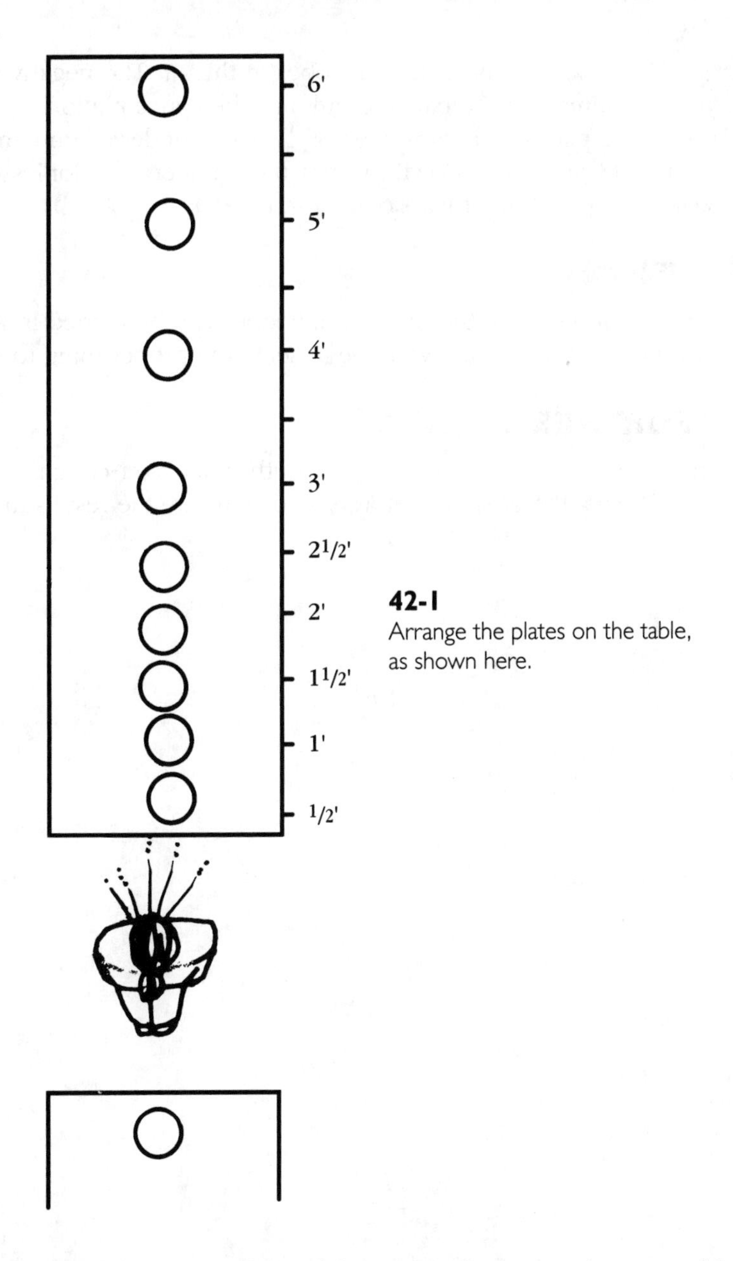

42-1
Arrange the plates on the table, as shown here.

Now stand at the end of the 6-foot table (between the two tables, as in Fig. 42-1), facing the long table. Have two helpers stand along the edge of the 6-foot table, wearing surgical face masks so they aren't infected with your germs and they don't inoculate the plates.

Have your helpers quickly remove all the plate lids. You should then cough immediately in the direction of the plates. Have your helpers wait 30 seconds, then replace the lids. Seal all the plates with adhesive tape along their edges. Wait a few minutes, then open the control plate up for 30 seconds, but don't cough into it. There are always some microbes in the air. The negative control will give you an estimate of the background microbe concentration.

Incubate all the plates in a warm area of the room under a lamp and check at 24 hours and 48 hours for growth. Count the number of colonies on each plate. Repeat the experiment at least one more time.

CONCLUSIONS

What do you conclude about the distance microbes can be carried in a cough? Does covering your mouth when you cough seem the proper thing to do?

GOING FURTHER

Do the same experiment, but cover your mouth when you cough or sneeze. What happens? Use a throat spray before you cough. Are the results the same?

Part nine

Microbes & the environment

Too often microbes are thought of just as disease-causing organisms. In truth, however, microbes are essential to the natural world. Because they are so small, we don't notice them when we take a nature walk. But nature would be very different were it not for the microbes. Microbes recycle nutrients and elements, such as carbon, nitrogen, phosphorus, and sulfur.

Most of this recycling takes place in the soil. Soil fertility and our ability to grow crops depends on microbes. The type and condition of the soil determines what type of microbes can grow and survive. The amount of water, oxygen, organic matter, and inorganic matter, plus the pH, all control what microbes will and won't grow. Temperature is also important. Some of the projects in this section investigate soil microbes.

Water microbes include consumers, producers, and decomposers, and are usually part of complex food webs. Freshwater, brackish waters, and salt water all contain microbes. Water often carries disease-causing microbes (*pathogens*). Many bodies of water have sewage directly dumped into it containing countless pathogens, so drinking water must be treated before it is safe to use. That is why chlorine is added to water supplies.

Algae, living in the oceans, create much of the world's oxygen supply during photosynthesis and are the first link in most aquatic food chains. This section contains a number of projects that explore microbes that live in the water.

Sewage is composed of human and animal waste. It has a high content of organic matter. This organic matter is an ideal food for microorganisms that decompose waste. You can also find microbes in sewage that normally live in our digestive tract. These can be dangerous if they are ingested, which is why it is important to have a clean water supply. This section also includes projects that investigate microbes and sewage.

Unlike water and soil, air is not an environment in which microbes grow. However, they are transported by the air. Some pathogens are transported through the air to get to their new hosts. Many molds depend on air to transport their spores to new homes. In energy-efficient buildings, with little air turnover, air can carry a higher number of microbes. This is especially a problem when the air in the building is moist. Air can be sampled easily for microbes by simply opening an agar plate and letting the microbes settle onto the plate. This section also contains projects that study microbes in the air.

43

Look, but don't touch

Nitrogen fixers

The air we breath consists of 78% nitrogen. *Nitrogen* is essential to almost all forms of life, but the nitrogen found in the air cannot be used by organisms. Certain types of bacteria, called *nitrogen fixers*, are able to convert the nitrogen in the air into a form that plants can use. Without these bacteria, life on our planet simply could not exist as we know it.

Legumes are plants that often live in a symbiotic relationship with nitrogen fixing bacteria. What is the effect of a nitrogen fixer on the growth of a legume?

MATERIALS

- Marker that writes on glass
- Six pots for plants
- Milled sphagnum moss (available from a nursery or garden store) or use potting soil
- Water
- Bean sprouts
- *Rhizobium leguminosarum* culture (available from scientific supply house)
- Sterile inoculating loops (follow procedures in chapter 1 "An Introduction to Microbiology" on how to sterilize inoculating loops)
- Magnifying glass or stereoscope
- Drawing paper and pencil

PROCEDURES

Label three of the pots "With" and the other three "Without." Fill all six pots with the milled sphagnum moss. Wet the moss thoroughly, then let the excess water drain out. Plant six bean sprouts in each pot. Place all the pots in a warm area with indirect sunlight. Keep watering the plants so they don't dry out.

After 5 days, inoculate the "with" pots with *Rhizobium leguminosarum* culture. Do this by sticking a sterile inoculating loop into the *Rhizobium leguminosarum* culture and then into the pot (see Fig. 43-1). Repeat this inoculation procedure three times for each "With" pot. If the *Rhizobium leguminosarum* culture is in a dry form, simply sprinkle it over the top of the moss and water. Grow the beans for 8 to 12 weeks. Keep watering them so they don't dry out. Make observations and take notes every few days. At the end of the experiment, pull up the roots of one of the "with" plants and compare them to the roots of a "Without" plant.

43-1
Insert the inoculating loop containing the microbes in the plant pots labeled "with."

CONCLUSIONS

How did the growth of the plants with and without the nitrogen fixers differ? Are the differences obvious? Were differences found throughout the growth period or just during certain portions of it? How did the root systems differ both in size and structure? Look at the roots under a magnifying glass or through a stereoscope and draw pictures of both. What is the effect of the Rhizobium leguminosarum culture on bean growth?

GOING FURTHER

Collect your own *Rhizobium*. Go to a farm growing a legume, such as beans or clover, and pull up a plant by the roots. Be sure to get permission from the property owner first. Keep the roots moist. Bring them home or to the lab, crush the root nodules, and place them in a nitrogen-poor soil. Repeat the plant growth experiment, using this nodule-treated soil and the regular untreated soil. Are your results the same?

44

Green dirt

The presence of algae in soils

Soil is a complex mix of organic and inorganic material. It is familiar to us as the home where earthworms, millipedes, and insects live; where plants put down their roots; and where dead organisms are decomposed.

But is soil also the home for another type of microbe—the algae? *Algae* are producers, meaning they use light from the sun to make their own food just as green plants do. Therefore they have *chlorophyll*, a green pigment used to capture the energy from the sun. Most algae live in water; they are the grass of the seas, but do algae live in the soil also?

MATERIALS

- Six pint-sized mason jars with lids
- 100 ml measurer
- Marker that writes on glass
- 800 ml of sterile algae nutrient medium or make your own using Bold's or Bristol's medium (all are available from scientific supply house)
- 1 teaspoon measurer
- Two soil samples (e.g., from a lawn, forest, or garden)
- Aluminum foil

PROCEDURES

Determine the height to fill a mason jar with 100 ml of liquid. Mark that line on all six of the jars. Label three of the jars with "Soil 1" and the other three with "Soil 2." Sterilize the mason jars and their lids by following the instructions in chapter 1 "An introduction to microbiology." While the jars are still warm, fill them with the sterile nutrient algae medium, up to the 100 ml line. Add 1 teaspoon of soil type 1 into each of the three jars labeled "Soil 1." Add 1 teaspoon of soil type 2 into the other three jars. Close the jars tightly and shake very well. After shaking, open all the lids so the jars are not completely sealed.

Wrap one jar of each soil type with aluminum foil, so no light can enter (see Fig. 44-1). These will be the negative control jars because algae requires light to

44-1 One of each set of three jars should be wrapped in foil.

grow. If there is no light available, there should be no algae growth; there might, however, be fungal or bacterial growth. Place all the jars in an area with indirect sunlight. If the sun is direct, the jars will warm the water in the jars too much and everything will die. Every 2 to 3 days, observe the jars for green growth.

CONCLUSIONS

Compare the color of the growth in jars with aluminum foil to those without. Any differences? Do you see any signs of algal growth in the soil samples? Are there any differences between the two types of soils?

GOING FURTHER

Use a microscope to try to identify the algae found in the different types of soil? How do the types of algae found in the soil differ from other types of algae found in water?

45

We call it home
Microbes in different types of soil

Soil is much more than just dirt. It is a complex mixture of inorganic and organic matter. Soil holds nutrients and water, which are essential for plant growth. Organic matter is recycled in the soil. This recycling occurs when dead plant material is converted by bacteria, fungi, and other organisms into compounds that live plants can use to grow. These soil microbes are essential for plant growth. Do some soils have more microbes than others?

MATERIALS

- Marker that writes on glass
- 18 sterile nutrient agar plates (available from scientific supply house)
- Four sterile pint-sized mason jars with lids
- 100 ml graduated cylinder with 1 ml markings
- Sterile water
- Potting soil
- Compost (available from a garden center)
- Peat (available from a garden center)
- 15 sterile pipettes graduated in 0.1 and 1 ml (available from scientific supply house)
- 1-inch adhesive tape
- Incubator (or warm area in room under lamp)
- Thermometer
- Magnifying glass or stereoscope

PROCEDURES

Label each of three plates with one of the following labels: "Peat," "Compost," and "Potting Soil." Sterilize the three mason jars according to the instructions in chapter 1 "An Introduction to Microbiology." Label each of these jars with one of the following labels: "Peat," "Compost," and "Potting Soil." Then, add 100 ml of sterile water to each jar.

Next add 1 teaspoon of each type of soil to the proper jar. Put the lids on the jars and shake vigorously. Let the jars stand for 15 minutes. Shake the jars once again and then proceed. With a clean, sterile pipette (see chapter 1 "An Introduction to Microbiology" on how to sterilize pipettes), remove a small sample from each jar and drop 1 ml onto the proper plate.

After applying the sample, tilt the plate to spread the liquid evenly over the plate's surface. Seal the plates with adhesive tape all around the edges. Incubate all the plates in an incubator, if available, at 37 degrees C or in a warm area of the room under a lamp. Try not to let the temperature fluctuate more than 2 degrees. Count the number of colonies on the plates after 48 hours.

CONCLUSIONS

Observe the colonies that grow on the plates, first with your eyes and then with a magnifying glass or stereoscope. Do you see any differences in the number of microbes between these three types of soils?

GOING FURTHER

Do this experiment with different types of culture media, such as Sabouraud or tryptic soy. Different media allow different types of microbes to grow. Do you get different results with the same samples?

46

Light & dark

Microbial communities under different light conditions

Natural bodies of water are home to thousands of different microbes. All the components of a *food web* (e.g., producers, consumers, decomposers) can be found in natural waters. From hot springs to cold arctic pools, from the deepest depths of the oceans to the bottom of a clear mountain lake—aquatic microbes exist. Various microbes can be found in different types of water. How does light affect the growth of all aquatic microorganisms?

MATERIALS

- Natural water sample
- Sterile 16 oz. mayonnaise jar
- Four sterile pint-sized mason jars
- Aluminum foil
- Eyedropper
- Microscope slides
- Coverslips
- Microscope (preferably with an oil-immersion objective)
- Immersion oil
- Drawing paper and pencil
- Book to identify aquatic microbes (especially algae)

PROCEDURES

Collect a natural water sample (e.g., marine, fresh, brackish) in a sterile 16 oz. mayonnaise jar. Sterilize the mason jars according to the instructions in chapter 1 "An Introduction to Microbiology" on how to sterilize glassware. While jars are warm, but not hot, pour ¼ of the natural water sample into each jar. Wrap two of the jars completely around with foil, so no light can enter. Leave all the jars in indirect sun for 2 weeks. Do not place them in direct sunlight because the light will heat up the contents of the jars and kill most of the organisms.

At the end of the 2-week period, take notes about the appearance of the contents of each jar. Next, use an eyedropper to take a water drop from each sample and put it on a microscope slide. Cover each drop with a coverslip. Try not to trap any air bubbles under the coverslip. View the slide under a microscope, first at 100× magnification, then at 400×, and finally under oil immersion. Draw the types of microbes you see. Identify as many of the forms as you can using a resource book on aquatic microbes.

CONCLUSIONS

Does light have an effect on the types of microbes you found? Does light increase the number of all microbes or just some? Do some microbes actually prefer the dark? What do you conclude about the effect of light on the different aquatic microbes?

GOING FURTHER

Collect a water sample from a stream and perform a similar experiment. Do you find the same type and number of microbes as you did in this project? Keep all the cultures over a period of many weeks. Do the communities of microbes change over time?

47

Living scum

Biofilms & biofouling

Microbes in moist or wet habitats often adhere to a surface, forming a living sheet or film. This mass of microbes on a surface is known as a *biofilm.* For example, when we wake up in the morning, our teeth and gums are often covered with a biofilm that has grown there overnight. The sides of aquariums have a biofilm growing on them, and snails feed on these microbes. Almost any surface that is moist can get a biofilm growth. This sometimes causes problems. Biofilm can disturb water treatment plants and other industries that use water. Ships get biofilm and other organisms feeding on this film beneath the hull. This can interfere with the boats movement through the water. This is called *biofouling.* Industries use different chemical substances to control biofilm and biofouling. What can be used to reduce or eliminate biofilm growth?

MATERIALS

- Six pint-sized mason jars
- Aquarium or pond water
- Two or three desk lamps

Note: For the next four paints, you must read the labels carefully to determine what type of additives are included. You need a very small amount of each.

- White oil-based paint with an additive that controls mold growth
- White oil-based paint without an additive that controls mold growth
- White latex paint with an additive that controls mold growth
- White latex paint without an additive that controls mold growth
- Microscope slides
- Marker that writes on glass
- Paper towels
- Microscope (one that has an oil-immersion objective)
- Water
- Coverslips
- Book to identify aquatic microbes (especially algae)

PROCEDURES

Fill each mason jar ¾-full with aquarium or pond water. The jars need not be sterile. Leave all the jars under a desk lamp for 7 days. Don't keep the lamp too close to the jars or the water will overheat, killing the microbes. Add more water if the water level drops. There should be a rich growth of microbes in the mason jars. If the jars are not greenish in color, add more pond or aquarium water and wait another few days. You must see algae growth in the mason jars; that is, the water should have a greenish tinge, before continuing.

Apply a thin coat of one of the aforementioned paints to one side of each of four glass microscope slides (see Fig. 47-1). Label each slide. Let the slides air dry for 2 days. Keep them in a well-ventilated room because they might give off

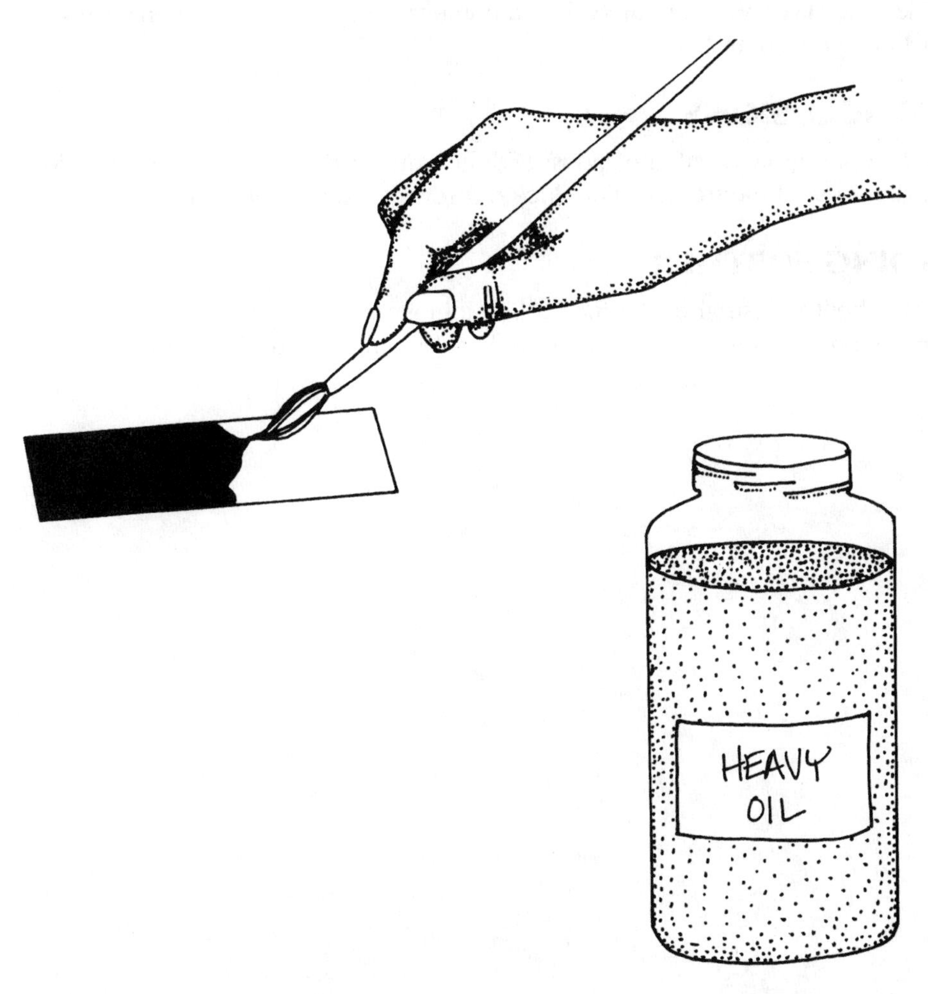

47-1 Apply a thin coat of each substance to one side of each slide.

harmful fumes. Place one slide in each mason jar by carefully dropping them into the water. Label each jar with the type of test material used on that slide. Put one uncoated slide in the fifth jar to act as a control. Leave all the slides in the jars for 14 days. During this period, add water to the jars to keep the slides underwater. Keep the jars under the lamp.

At the end of 2 weeks, remove all the slides. Be sure the labels are still visible on each slide. Wipe the "uncoated" side of the slides clean with the paper towels. Observe the slides under a microscope, at 100× and 400×. You could also look at the slides with a stereoscope. Make notes on the type and number of organisms found on each slide.

Finally, put a drop of water on the center of each slide and cover the drop with a coverslip. View each slide under 40×, 400×, and then an oil-immersion objective. Make notes on the type and number of organisms found on each slide including the control.

CONCLUSIONS

Did biofilm grow on all the slides? Did the same amount grow on all the slides or did some of the treatments reduce or prevent growth completely?

GOING FURTHER

Read about industrial or marine biofilm-inhibitors. Do they cause any environmental problems? Can you find examples of biofilms around or on your house?

48

A microbe high rise

Microbe communities in Winogradsky Columns

In the late 1800s, Sergei Winogradsky discovered that some bacteria use light from the sun to create their own food. Before that time, scientists believed that only plants and algae could do this. Winogradsky developed a miniature microbial world for the study of soil bacteria. He discovered that a sample of soil containing the correct substances, in the presence of light, would form a complex world of microbes. These miniature worlds are called *Winogradsky Columns.*

The bacteria in the soil at the bottom of the column change the environment in other portions of the column. This creates new environmental conditions for other bacteria to thrive. Make two Winogradsky Columns and use them to determine how fresh air at the bottom of a column can change the environment and affect the microbes living there.

Note: This is an advanced-level project.

MATERIALS

- One 16 oz. mayonnaise jar
- Mud from a lake or pond
- ½ cup measurer
- ¼ cup measurer
- Shredded newspaper
- Shredded leaves
- 1 teaspoon measurer
- Small bag of calcium carbonate ($CaCO_4$)
- Small bag plaster of Paris ($CaSO_4$)
- Long-handled wooden spoon
- 10 feet of ⅛-inch-diameter air hose (for aquarium air pump)
- Aquarium air pump
- Two 1-liter clear graduated cylinders (available from scientific supply house)

- 1300 ml of pond water
- Aluminum foil

PROCEDURES

Fill a 16 oz. mayonnaise jar with mud collected from a shallow pond or lake. Try to find mud that smells of rotten eggs because this will work best. Set aside ½ cup of this mud. Mix the remaining mud that you've collected with ¼ cup of finely shredded newspaper and ¼ cup of finely shredded leaves. Add 2 teaspoons of calcium carbonate and 2 teaspoons of plaster of Paris. Mix it all into the mud in the mayonnaise jar with the long-handled wooden spoon.

Take the air tube and connect it to an aquarium air pump. Insert the other end of the tube down inside one of the graduated cylinders so that the bottom of the tube is nearly at the bottom of the cylinder (see Fig. 48-1). Now tightly pack the mixture of mud, paper, and leaves into the bottom of both graduated cylinders (with and without the tube). Be careful not to trap any air bubbles in the mud layer. The packed layer of mud should extend from the bottom of the cylinder (1000 ml mark) to the 700 ml mark.

Layer a small amount of unchanged mud, which had been set aside at the start of the project, on top of the mixed mud in each cylinder. This should bring the mud layers up to the 650 ml mark in each cylinder. Very slowly add the pond water to fill each cylinder to the 0 ml mark, the top of the cylinder. Be careful not to disturb the mud layers when you pour the water into the cylinders. Cover both cylinders with foil.

Turn on the air pump at the low setting—you do not want a huge force disturbing the layers. Place both cylinders—one with and the other without the bubbler—in a north-facing window that gets indirect sunlight. Leave both cylinders in place for 6 weeks. Look for color changes in each cylinder each week. Watch for the formation of layers in each column.

CONCLUSIONS

Did both columns form layers? Can you determine what type of organisms are living in each layer and why? What is the effect of fresh air at the bottom of the tube? How does it affect the layers?

GOING FURTHER

Take samples from the layered column (using a long-stemmed Pasteur pipet) every week and view the samples microscopically at 400× and oil immersion. Note the change in the different layers of the column with time. What type of organisms are living in this column that are not living in the other column at the same level?

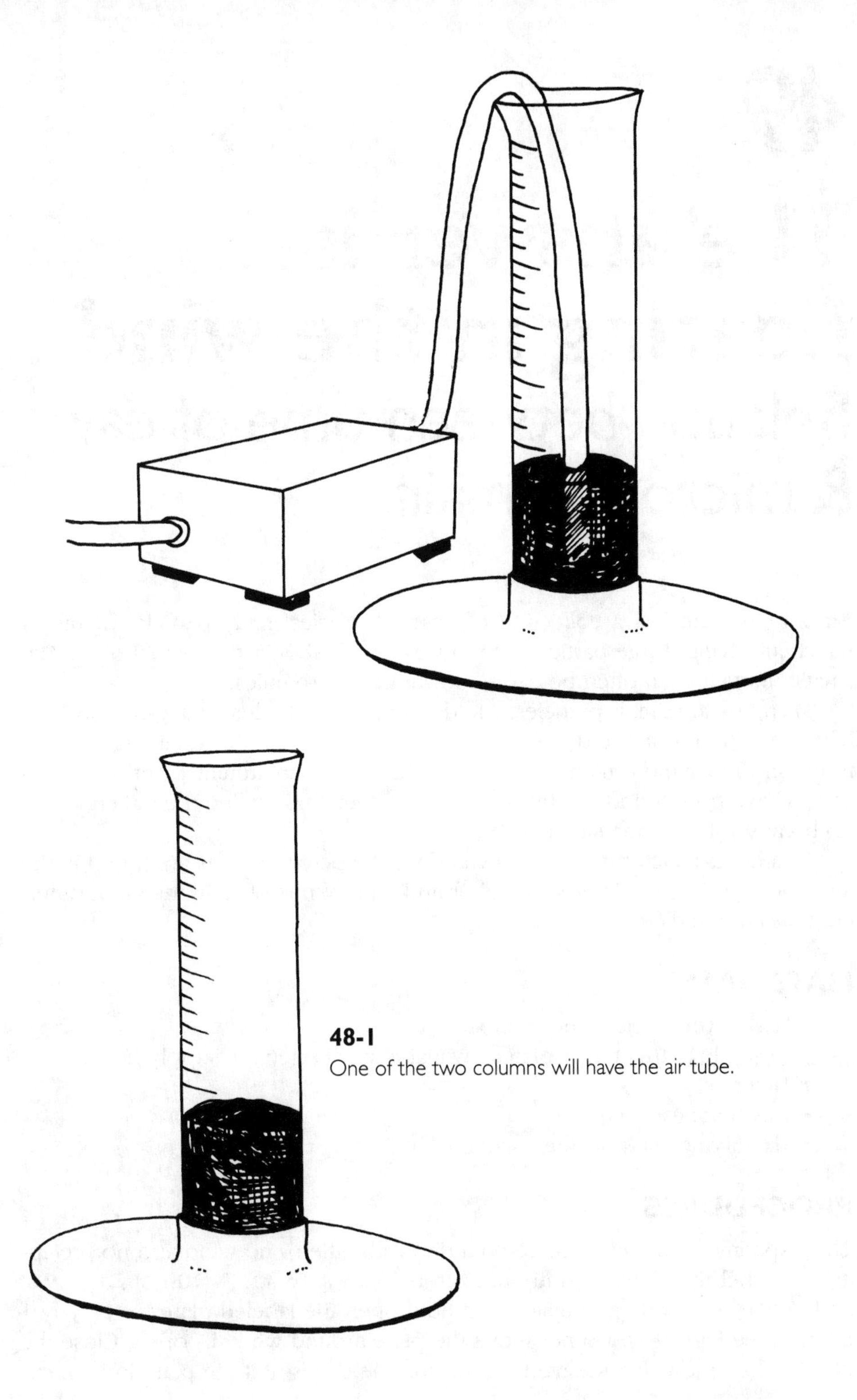

48-1
One of the two columns will have the air tube.

49

The answer is floating in the wind

Relation between time of day & microbes in air

Air is more than just a collection of gases. It carries many particles, some of which are living. Large particles fall out of the air sooner than small ones. The smaller particles can often be carried for hundreds of miles.

Microbes are small particles. Mold spores from molds that grow on land have been found hundreds of miles out to sea. Sewage bacterial cells can be found in the air more than 100 feet from a sewage treatment plant. Microbial spores have been found at altitudes of 90,000 feet! People find their allergies act up because of the spores carried by air.

Weather is a factor in how many and what type of particles are found in the air. Would you find a different number and type of microbes in the air at different times of the day?

MATERIALS

- Marker (that writes on plastic)
- 24 sterile nutrient agar plates (available from scientific supply house)
- Timer
- 1-inch adhesive tape
- Magnifying glass or stereoscope

PROCEDURES

This experiment should be done on a day with little or no wind and no precipitation. Label the plates of nutrient agar as follows: "6 A.M.," "10 A.M.," "2 P.M.," and "6 P.M." At 6 A.M., go outside and hold open the labeled plate for exactly 1 minute (see Fig. 49-1). Do not move the plate around while its open. Close the plate and seal it with adhesive tape around the edges. Put the plate in an incu-

49-1 Have a clock with a second hand available as you hold open the plate.

bator or in a warm area in the room under a lamp so that the temperature of the plate is approximately 30 degrees C.

Repeat this procedure at 10 A.M., 2 P.M., and 6 P.M. Observe the sealed plates at 24 hours and 48 hours after they were exposed to the air. Because they were opened at different times, you will be making these observations at different times. Note the number and shape of the colonies that form. Repeat this procedure on five different days. There should be no wind or rain on any of the days. Look at the colonies with a magnifying glass or stereoscope to see more detail.

CONCLUSIONS

Compare the growth on the plates at various times of day and on different days. Does it appear that there are different amounts and types of microbes floating in the air at different times of day and on different days? What is the reason for these differences?

GOING FURTHER

Repeat this with a different culture medium, such as Sabouraud media. Do you get the same results? Repeat this at a different time of the year. Did you get the same results?

50

Pasteur takes a ride

Wind speed & microbes

Louis Pasteur carried out an experiment about microbes in air. In his experiment, as he traveled through France, at each stop, he opened a sterile vial containing nutrient fluid. He theorized that all air contains microbes, but at higher altitudes, and because air is less dense, there would be fewer microbes. He traveled in the lowlands and up mountains. He found that growth appeared fastest in tubes opened at the lower elevations, compared with tubes opened in the mountains.

Today people travel primarily by car. Conduct a modern day Pasteur journey, but rather than determine how the number of microbes changes with altitude, see how they change with wind speed using an automobile. Does the speed of travel affect the amount of microbes collected?

MATERIALS

- Marker (that writes on glass)
- 15 sterile test tubes with nutrient broth (available from scientific supply house)
- Car with legal driver
- Access to a side street with little or no traffic
- Timer
- Tube rack

PROCEDURES

Label three sets of tubes with each of the following: "0," "5," "15," "30," and "50." These represent the speed the car is traveling when the vials are opened. Sit in the back seat of the car on the side away from oncoming traffic (see Fig. 50-1). Open the window. Ask the driver to start the car, but not to begin driving. Open a tube marked "0," stick it out the window, count to 3, and then recap the tube. Repeat for the other two tubes labeled "0." Note the time and then put the tubes back into the tube rack.

Warning: There is no need to hold your hand out of the car window more than a few inches. Be aware of oncoming objects along the roadside. Ask the driver to drive 5 miles per hour. While the car is moving, open a tube marked

50-1 Hold the open vial out the window of the car at a 45-degree angle.

"5," stick it out the window, hold it at a 45-degree angle, with the open end pointing forward. Count to 3 and recap the tube. Put this tube back in the rack. Repeat this for the other two "5" tubes. Note the time. Repeat this at 15 miles per hour and 30 miles per hour.

Return to the lab or your home lab and observe the tubes every 15 to 20 minutes. Gently tap the tubes to distribute any growing microorganisms. Continue to observe the tubes every hour. Note the time when you first observed cloudiness in each tube. The time it took for the first visible growth to appear is a rough estimate of the number of microbes that were captured when the vials were opened (inoculated).

CONCLUSIONS

What do you conclude about the number of microbes inoculating (contaminating) a nutrient vial at different car speeds? Does the speed of travel affect the size of the original inoculum? Do all the tubes opened at the same speed have growth beginning at the same time?

GOING FURTHER

Research whether the results of your experiment mean that people living in windy areas are more prone to catching other people's germs. Investigate whether colder climates contain fewer airborne microbes than warmer climates.

51

Water, water everywhere, hardly a drop to drink

Survey of bacteria from different water samples

A supply of clean water is crucial to our continued health and survival. Clean water has fewer microbes than dirty water and will also not have any disease-carrying organisms(pathogens). If *coliform bacteria* are found in water supplies then water is considered *dirty*. Coliform bacteria get into the water supply when untreated sewage is mixed with the water because human and animal wastes usually contain coliform bacteria.

Towns frequently treat their water supplies with chlorine, to kill the coliform bacteria and other potential disease-causing organisms. Various natural waters can contain different numbers and types of microbes. How many types of bacteria and other microbes and how many of each exist in different types of water?

MATERIALS

- 1 cup measurer
- Samples from any four of the following water sources:
- Water from a pond (with lots of green algae)
- Water from an aquarium
- Water from a stream
- Water from a puddle
- Swimming pool water
- Tap water
- Bottled water
- Four sterile pint-sized mason jars with lids

- 10 sterile nutrient agar plates (available from scientific supply house)
- Five sterile pipettes graduated 0.1 and 0.5 ml (available from scientific supply house)
- 1-inch-wide adhesive tape
- Incubator (or warm area in room under lamp)
- Thermometer
- Magnifying glass or stereoscope
- Paper and pencil

PROCEDURES

Collect 1 cup of water from the different sources listed above. For each water source, label two sterile nutrient agar plates with the name of the source. Mark one of each plate "0.1" and the other "0.5." This represents the amount of water applied to the plate. Use a pipette to inoculate 0.1 ml of the water sample on the properly labeled plate (see Fig. 51-1). Close the plate and rotate it to spread the sample equally over the plate's surface. Repeat this procedure using a .5 ml sample on the .5 ml plate.

51-1
Inoculate each plate with a pipette.

Repeat this entire procedure for each water source. Seal all the plates with adhesive tape. You should have four sets of two plates. Incubate all the plates in an incubator at 37 degrees C or in a warm area of the room under a lamp. Use a thermometer to be sure the temperature is correct and doesn't fluctuate more than 2 degrees. After 48 hours, observe all the plates and count the number of bacteria colonies found on each plate. Use a magnifying glass or stereoscope to see the colonies in more detail.

CONCLUSIONS

Graph the number of colonies per plate for each source of water for the 0.1 ml plates and for the 0.5 ml plates. Do some water sources have far more bacteria than others? Do all have bacteria? Are there five times as many bacteria on the 0.5 ml plates as on the 0.1 ml plates? What do you conclude about the bacterial content of different water sources?

GOING FURTHER

How many different types of bacteria are found in the different water sources? Use a microscope with oil immersion to identify these bacteria. Use different media, such as tryptone agar. Do you get different results? Investigate various types of water pollution that involve all microbes or just bacteria, and study how we try to control them.

A
Using metrics

Most science fairs require all measurements be taken using the metric system as opposed to English units. Meters and grams, which are based on powers of 10, are actually far easier to use during your experimentation than feet and pounds. You can convert any English units into metric units, if need be, but it is easier to simply begin with metric units. If you are using school equipment, such as flasks or cylinders, check the markings to see if any use metric units. If you are purchasing your glass or plastic ware, be sure to order metric markings.

Conversions from English units to metric units are given below, along with frequently used abbreviations. All conversions are approximations.

Length
1 inch (in.) = 2.54 centimeters (cm.)
1 foot (ft.) = 30.4 cm.
1 yard (yd.) = .90 meters (m.)
1 mile (mi.) = 1.6 kilometers (km.)

Volume
1 teaspoon (tsp.) = 5 milliliters (ml.)
1 tablespoon (tbsp.) = 15 ml.
1 fluid ounce (fl. oz.) = 30 ml.
1 cup (C.) = .24 liters (l.)
1 pint (pt.) = .47 l.
1 quart (qt.) = .95 l.
1 gallon (gal.) = 3.80 l.

Mass
1 ounce (oz.) = 28.00 grams (g.)
1 pound (lb.) = .45 kilograms (kg.)

Temperature
32 degrees Fahrenheit (F) = 0 degrees Celsius (C)
212 degrees F = 100 degrees C

B

Scientific supply houses

You can order equipment, supplies, and live specimens for projects in this book from these companies. For your convenience, a list of catalog order numbers from Ward's Scientific Supply is listed in Appendix C, categorized by project number.

Blue Spruce Biological Supply Company
221 South Street
Castle Rock, Colorado 80104
(800)621-8385

The Carolina Biological Supply Company
2700 York Road
Burlington, North Carolina 27215
Eastern US: 800-334-5551
Western US: 800-547-1733

Connecticut Valley Biological
82 Valley Road
P.O. Box 326
Southampton, Massachusetts 01073

Fisher Scientific
4901 W. LeMoyne Street
Chicago, Illinois 60651
(800)955-1177

Frey Scientific Company
905 Hickory Lane
P.O. Box 8101
Mansfield, Ohio 44901
(800)225-FREY

Nasco
901 Janesville Avenue
P.O. Box 901
Fort Atkinson, Wisconsin 53538
(800)558-9595

Nebraska Scientific
3823 Leavenworth Street
Omaha, Nebraska 68105
(800)228-7117

Powell Laboratories Division
19355 McLoughlin Boulevard
Gladstone, Oregon 97027
(800)547-1733

Sargent-Welch Scientific Company
P.O. Box 1026
Skokie, Illinois 60076

Southern Biological Supply Company
P.O. Box 368
McKenzie, Tennessee 38201
(800)748-8735

Ward's Natural Science Establishment, Inc.
5100 West Henrietta Road
Rochester, New York 14692
(800)962-2660
815 Fiero Lane
P.O. Box 5010
San Luis Obispo, California 93403
(800)872-7289
(Order numbers from Ward's catalog are provided in Appendix C)

C
Ward's scientific supply catalog numbers

The following items (listed in the Materials sections of the projects) can be purchased from Ward's Natural Science Establishment, Inc. The catalog numbers for these materials are listed below.

Blood agar plates (88W0901)
Sabouraud-dextrose agar plates (88W0920)
Tryptic Soy agar plates (88W0925)
Tryptic soy broth tubes (88W0819)
Nutrient agar media to make agar plates (in premeasured, Media Pour Paks) (88W0004)
Litmus milk (38W0366)
Antibiotic disks (38W1601 among others)
Mixed protist culture (87W1510)
Disposable sterile inoculating loops (14W0954)
Sterile swab applicators (14W5502)
India ink (15W9843)
Autoclavable Biohazard Bags (18W6905)

Glossary

aerobic Being in the presence of oxygen.
agar Gelatin like substance from seaweed; used as solid growth media for microbial cultures.
algae Single-celled microscopic organisms that contain chlorophyll and photosynthesize; some live in large colonies or consist of long filaments and are, therefore, macroscopic.
anaerobic Living in the absence of oxygen.
autoclave A machine that sterilizes objects by combining moist heat (steam) and pressure.
autotroph Organism able to produce their own food (by photosynthesis using the sun or from inorganic chemical energy).
bacteria A single-celled microscopic organism that reproduces by fission and has no nuclear membrane.
biochemistry The study of biochemical reactions.
biofilm A sheet of microbial growth present on a surface.
catalyst A chemical that helps a reaction go to completion but is itself not used up by the chemical reaction.
cell The basic unit of life; all cells are bags containing liquid (the cytoplasm); the bag itself is the cell membrane.
cilia Tiny, short hair-like structures used by some organisms to move.
ciliate A type of protozoa with cilia.
colony A population of cells growing on a solid medium.
consumer An organism that eats producers or other consumers.
culture [noun] A colony of microbes with all ingredients necessary for their survival; [verb] to grow microbes.
decomposer Organism able to break down dead organic material such as the dead bodies of animals or dead plant leaves; Fungi and many bacteria are decomposers.
enzyme A protein that catalyzes (helps) a biochemical reaction.
eukaryotic A type of organism whose cells have internal organelles and internal membranes, such as a nucleus; all nonbacterial organisms are eukaryotic, including the higher plants and animals.
exoenzyme An enzyme that is excreted outside of a cell.
filter sterilization Forcing a liquid through a filter with pores less than 0.20 microns; The liquid is sterilized because the microbes cannot fit through the small pores while the liquid passes through.
flagellates A type of the protozoa that use flagellum to move.
flagellum A tail-like structure used by microorganisms to move.

fungi A group of organisms that are heterotrophic (consume their food) and have a cell wall; Most are decomposers.

germinate To begin to grow.

hemolytic Having the ability to breakdown (lyse) blood cells.

heterotroph Organisms that require an external source of organic chemical energy (food) to survive (as opposed to autotrophic).

host The organism that supports the life of a parasite (e.g., humans are the host for cold viruses).

hypha The filaments of a mold or a mushroom; In molds, the tip of the hyphae produce spores that produce the next generation.

infection A growth of microorganisms within a host.

inoculum The starting material for a microbial culture.

lyse To break open.

macroscopic Large enough to see with the unaided eye.

metabolism The sum of chemical and biochemical reactions necessary for life.

microbe A small organism visible only with a microscope; could be a bacteria, alga, fungus, protist, or virus.

microscopic Not visible with the unaided eye.

mold A general term for many of the simple, filamentous fungi.

morphology The study of the appearance of an organism, including its shape, texture, and color.

mycelium Main body of a multicellular fungus.

mycosis Infectious diseases caused by a fungus.

nucleus A membrane-enclosed structure that contains genetic material in a eukaryotic cell.

obligate pathogens Microorganisms that must have a host to survive and reproduce; In comparison to facultative, pathogens are accidental contaminants of a host and can survive outside of a host.

organelle A membrane-enclosed structure within a cell in eukaryotic organisms.

parasite An organism that can live on or in another organism.

Pasteur Early microbiologist who discovered methods to reduce microbial growth in foods (pasteurization) and discovered the rabies vaccine, among other things.

petri plate A shallow plate or dish used to hold a usually solid growth medium to culture microbes.

pipette A long, thin tube that holds a known amount of liquid; used to transfer liquids.

producer An organism that makes its own chemical energy (food), usually from the energy in the sun (but some bacteria are producers using the energy in inorganic molecules).

protists A kingdom of living things, composed primarily of single-celled organisms that do not have a cell wall; some have chlorophyll, while others do not.

protozoa A group of protists that do not contain chlorophyll; complex, single-celled eukaryotes with a cell membrane.

resolving power The smallest distance between two objects at which the objects can still be seen as separate things; below this distance the two objects appear as one.

selective media Media that contains one or more chemicals that do not permit some microbes to grow, but do permit the growth of the microbe of interest.

species Organisms with the potential to breed and produce viable offspring.

spore-forming bacteria Prokaryotes that produce spores that are resistant to drying and can survive difficult environmental conditions.

sterile Absence of life.

viable Able to live.

virus A package of genetic material surrounded by a protein capsule that requires a living host to reproduce.

Bibliography

Anderson, D.A., and R.J. Sobieski. 1980. *Introduction to microbiology.* 2nd ed. St. Louis, MS: The C.V. Mosby Company.

Bleifeld, M. 1988. *Experimenting with a microscope.* New York: Watts.

Jahn, T.L., E.C. Bovee, and F.F. Jahn. 1979. *How to know the protozoa.* 2nd ed. Dubuque, IO: Wm. C. Brown Publ. Co.

Lang, S. 1992. *Invisible bugs and other creepy creatures that live with you.* New York: Sterling Publishers.

Nardo, D. 1991. *Germs: Mysterious microorganisms.* San Diego: Lucent Books.

Oxlade, C., and C. Stockley. 1989. *The world of a microscope.* Tulsa, OK: EDC Publishers.

Pelczar, M.J., and R. Reid. 1965. *Microbiology.* 2nd ed. New York: McGraw-Hill.

Poindexter, J. 1971. *Microbiology: An introduction to protists.* New York: MacMillan.

Prescott, G.W. 1978. *How to know the freshwater algae.* 3rd ed. Dubuque, IO: Wm. C. Brown Publ. Co.

Reid, T. K. 1967. *Pond life.* New York: Golden Press.

Sabin, F. 1985. *Microbes & bacterias.* Mahwah, NJ: Troll Assocs.

Stewart, G. 1992. *Microscopes: Bringing the unseen world into focus.* San Diego: Lucent Books.

Tortora, G. 1989. *Microbiology: An introduction.* Redwood City, CA: Benjamin Cummings.

Index

A

B

C

D

E

F

G

H

I

N

O

P

About the author

The author is an adjunct professor of environmental science at Marymount College in Tarrytown, New York. He is the founder of the Center for Environmental Literacy, which was created to educate the public and business community about environmental topics. He holds a B.S. in biology and an M.S. in entomology. He is the author of many books that simplify science and technology.